Monographs in

Volume 6

Development of Elliptic Functions According to Ramanujan

Monographs in Number Theory ISSN 1793-8341

Published

Vol. 1 Analytic Number Theory: An Introductory Course
by Paul T. Bateman & Harold G. Diamond

Vol. 2 Topics in Number Theory
by Minking Eie

Vol. 3 Analytic Number Theory for Undergraduates
by Heng Huat Chan

Vol. 4 Number Theory: An Elementary Introduction Through Diophantine Problems
by Daniel Duverney

Vol. 5 Hecke's Theory of Modular Forms and Dirichlet Series (2nd Printing with Revisions)
by Bruce C. Berndt & Marvin I. Knopp

Vol. 6 Development of Elliptic Functions According to Ramanujan
by K. Venkatachaliengar, edited and revised by Shaun Cooper

Monographs in Number Theory

Volume 6

Development of Elliptic Functions According to Ramanujan

originally by

K Venkatachaliengar

edited and revised by

Shaun Cooper

Massey University, New Zealand

NEW JERSEY • LONDON • SINGAPORE • BEIJING • SHANGHAI • HONG KONG • TAIPEI • CHENNAI

Published by

World Scientific Publishing Co. Pte. Ltd.
5 Toh Tuck Link, Singapore 596224
USA office: 27 Warren Street, Suite 401-402, Hackensack, NJ 07601
UK office: 57 Shelton Street, Covent Garden, London WC2H 9HE

British Library Cataloguing-in-Publication Data
A catalogue record for this book is available from the British Library.

Monographs in Number Theory — Vol. 6
DEVELOPMENT OF ELLIPTIC FUNCTIONS ACCORDING TO RAMANUJAN

ISBN-13 978-981-4366-45-8
ISBN-10 981-4366-45-5

Printed in Singapore.

This book is dedicated to the memory of V. Ramaswami Iyer, the founder of the Indian Mathematical Society (I.M.S); Prof. A. Narasinga Rao, one of the earliest members of the I.M.S.; and my revered teacher Dr. B. S. Madhava Rao of the University of Mysore who passed away recently at Bangalore. I may say that these are the noblest academic people I have known in my life.

K. Venkatachaliengar

Preface

Preface to the original work

This book is devoted to the development of elliptic functions as perceived by Srinivasa Ramanujan. Ramanujan had not seen any standard book on elliptic functions before he went to England. This is clear since he did not recognize their characteristic properties—double periodicity, addition theorem, etc.,—and the notation he uses is his own. Nowhere in his treatment do we find the familiar Legendrian k, k', K, K', E, E', or the Jacobian parameter q, or the Weierstrassian $\wp$, ζ, σ, g_2, g_3, e_1, e_2 or e_3. He uses α for k^2; z for K; Q and R instead of g_2 and g_3; and so on. It is only when he published in England his two papers: "Modular equations and approximations to π" [90], and "On certain arithmetical functions" [91], he uses the modern notations and expresses his parameters in terms of the classical ones. He proves his basic identity in [91, (17)] (see (1.5)), modestly claims that the results developed there really do belong to the theory of elliptic functions, and draws the interest of the reader to the simplicity of his proofs. One finds in his paper [91] the simplest proof of the factorization of the discriminant. First, if

$$P = 1 - 24\sum_{j=1}^{\infty}\frac{jq^j}{1-q^j}, \quad Q = 1 + 240\sum_{j=1}^{\infty}\frac{j^3q^j}{1-q^j}, \quad R = 1 - 504\sum_{j=1}^{\infty}\frac{j^5q^j}{1-q^j}$$

and $|q| < 1$, then we have Ramanujan's differential equations:

$$q\frac{dP}{dq} = \frac{P^2 - Q}{12}, \qquad q\frac{dQ}{dq} = \frac{PQ - R}{3}, \qquad \text{and} \qquad q\frac{dR}{dq} = \frac{PR - Q^2}{2}.$$

Ramanujan used these differential equations to derive the factorization formula:

$$Q^3 - R^2 = 1728q\prod_{j=1}^{\infty}(1-q^j)^{24}.$$

The proof of Ramanujan's differential equations is by use of the basic Ramanujan identity (1.5) which is entirely algebraic in nature. Ramanujan's entire development is also algebraic in character.

A generalized form of Ramanujan's identity—see (1.9) and (1.13)—can be guessed from his work and its proof is almost immediate. This enables us to derive the addition theorem and differential equation of the classical function $\wp(\theta)$ of Weierstrass as well as the proof of Ramanujan's differential equations mentioned above. A further generalization—see (3.17)—yields the corresponding results for the Jacobian elliptic functions.

As is well known, the theory of elliptic functions contains proofs of some famous and beautiful identities. All of these can be derived in a simple way following the work of Ramanujan.

The most intricate part of Ramanujan's work in this theory is the modular equations given in various forms, and the evaluation of the corresponding singular moduli. Ramanujan has not indicated the proofs of these, especially for the latter; only bare outlines of the findings of modular equations of lower degree are sketched in his notebooks. In Chapter 7, we will give proofs of the modular equations of degrees 3, 5, 7, 11 and 23 based solely on the quadratic transformation of Legendre's modular function λ.

Dr. V. R. Thiruvenkatachar, my lifelong friend and colleague, has cooperated with me in the entire work concerning this book. My sincere thanks to him.

I am grateful to the authorities of the Madurai Kamaraj University, especially the Vice Chancellor Prof. S. Krishnaswamy, Professors K. R. Nagarajan and T. Soundararajan, and Drs. T. V. S. Jagannathan and R. Bhaskaran, who have spared no effort to bring out the book in a very nice form.

I hope that the publication of this book will revive interest in the study of the theory of elliptic functions in the Ramanujan way. Many of the results given here with simple proofs can be used in the teaching of function theory.

K. Venkatachaliengar

Preface to the revised version

The monograph "Development of Elliptic Functions according to Ramanujan", by K. Venkatachaliengar [105] was originally published as Technical Report 2 by Madurai Kamaraj University in February, 1988. In a letter to the author, A. Weil wrote [109]:

I can well appreciate the difficulty and value of writing such a book, and I am sure the mathematical world will be grateful to you for having written it. ... it clearly reflects much work and a deep familiarity with the subject, and the points you make in connection with Ramanujan seem to me entirely convincing.

In his review of the monograph, B. C. Berndt remarked [13]:

The author has studied Ramanujan's papers and notebooks over a period of several decades. His keen insights, beautiful new theorems, and elegant proofs presented in this monograph will enrich readers ....

The views expressed by Weil and Berndt are shared by many mathematicians, so it is worthwhile to publish a corrected and edited version of the monograph with complete proofs and references.

My philosophy in preparing the revised version has been to follow Venkatachaliengar's methods and to preserve the ideas from the original monograph as much as possible. Where gaps have been found, every effort has been made to complete the proof. Occasionally it has been necessary to give an alternative proof along different lines. To bring the work up to date, additional comments have been included in notes at the end of each chapter. References, both to 19th century works as well as to recent publications, have been vigorously sought and recorded in the bibliography.

The first three chapters as far as Sec. 3.3 follow the original monograph closely. Several details have had to be worked out in order to complete the proofs in Sec. 3.4. Perhaps the biggest sticky point in the book is the unmotivated definition of x and z by (3.57) and (3.58): the proofs are correct, but the formulas must be known in advance. Even with this disadvantage, the development that follows is interesting and deserves to be studied. Some comments that may be used to motivate these formulas have been provided in the notes at the end of Chapter 3. Each of Secs. 5.4–5.6 begin with formulas that must be known in advance. References to other methods are given in the notes at the end of Chapter 5.

Another issue that required a lot of attention was the reconciliation of the results from Secs. 3.4 and 4.4 that occurs in Sec. 5.1. The reader is referred to the notes at the end of Chapter 5 for further references to alternative methods.

The title of Chapter 7 was originally "Picard's Theorem", but the chapter is really about the modular function λ, so it has been renamed. Some of the material in Sec. 7.1 has been reworked completely. I could not fix Venkatachaliengar's proof that $|a_n| < 41^n$, so I used similar ideas to give a different proof of the stronger result $|a_n| \leq 16^n$ given by (7.8), instead. It is well-known that the properties of the modular function λ can be used to prove Picard's theorem. The details given by Venkatachaliengar in [105, pp. 120–123] rely on selectively choosing branches of the square-root function to erroneously deduce a mapping property of an analytic function. This result could not be fixed, so a proof of Picard's theorem is not given in this revised version. Instead, the reader is referred to Venkatachaliengar's paper [103] for another proof of Picard's theorem, or to one of the many texts on complex analysis for the standard proof. For example, a clear and succinct deduction of Picard's theorem from properties of the modular function λ is given as an exercise in [29, p. 119].

The original monograph [105] has been reviewed by Berndt [13] and Cooper [48]. Parts of the monograph have been surveyed and extended in the works of S. Bhargava [27], Cooper [44] and Venkatachaliengar [106].

The following notation will be used throughout. Let τ be a complex number with positive imaginary part, and let h and q be defined by

$$h = \exp(i\pi\tau) \quad \text{and} \quad q = \exp(2\pi i\tau)$$

so that $q = h^2$.

In the early 1990's my teacher R. A. Askey gave me a copy of Venkatachaliengar's monograph and suggested I edit it. B. C. Berndt encouraged me to complete the editing, and always enquired as to how it was going whenever we met at a conference. Venkatachaliengar's former Ph.D. student V. Ramamani kindly took me by auto-rickshaw to meet Venkatachaliengar at his house in Bangalore on July 4, 2000. Perhaps the most ardent supporter of Venkatachaliengar's work is his grandson G. N. Srinivasa Prasanna. The conference that Prasanna organized in Bangalore as a tribute to the life and work of Venkatachaliengar during July 1–5, 2009, provided the final stimulus to complete the editing of the monograph. The conference proceedings [8] contain information about Venkatachaliengar's life and work. Berndt and H. H. Chan offered valuable comments on a draft of the revised manuscript. To all of the people mentioned in this paragraph, I offer my heartfelt thanks.

Shaun Cooper, Auckland, April 2011

Contents

Chapter 1

The Basic Identity

1.1 Introduction

Ramanujan has indicated how one can, in a simple, purely algebraic way, develop the basic properties of the classical elliptic and other allied functions. Professor Birch [28] of the University of Oxford considers Ramanujan's paper [91] to be *"one of the most beautiful papers published by this (Cambridge Philosophical) Society"*.

The title of Ramanujan's paper—"On Certain Arithmetical Functions" —does not reveal the rich and manifold contents. The first three sections of the paper contain a few asymptotic formulas concerning the divisor sums $\sigma_r(n) = \sum_{d|n} d^r$ and a connected convolution function. Sections 4–10 deal with elliptic functions, more particularly the Weierstrass elliptic function expressed by its Fourier development. In Sec. 11, Ramanujan obtains closed formulas for several connected convolution functions. Section 12 contains the beautiful identity [91, (59)]:

$$\frac{1^5 q}{1-q} + \frac{3^5 q^2}{1-q^3} + \frac{5^5 q^3}{1-q^5} + \cdots = Q \times \left(\frac{q}{1-q} + \frac{3q^2}{1-q^3} + \frac{5q^3}{1-q^5} + \cdots \right)$$

where

$$Q = 1 + 240 \left(\frac{1^3 q}{1-q} + \frac{2^3 q^2}{1-q^2} + \frac{3^3 q^3}{1-q^3} + \cdots \right),$$

as well as a nice deduction of a convolution property for divisor sums. Here, and throughout this book, q is a complex variable that satisfies $|q| < 1$.

Let

$$f(q) = q^{1/24}(1-q)(1-q^2)(1-q^3)\cdots.$$

Ramanujan determines orders of several arithmetical functions using the four interesting formulas [91, (65)]:

$$f(q) = q^{1^2/24} - q^{5^2/24} - q^{7^2/24} + q^{11^2/24} + \cdots, \tag{1.1}$$

$$f^3(q) = q^{1^2/8} - 3q^{3^2/8} + 5q^{5^2/8} - 7q^{7^2/8} + \cdots, \tag{1.2}$$

$$\frac{f^5(q)}{f^2(q^2)} = q^{1^2/24} - 5q^{5^2/24} + 7q^{7^2/24} - 11q^{11^2/24} + \cdots, \tag{1.3}$$

$$\frac{f^5(q^2)}{f^2(-q)} = q^{1^2/3} - 2q^{2^2/3} + 4q^{4^2/3} - 5q^{5^2/3} + \cdots, \tag{1.4}$$

where 1, 5, 7, 11, ..., are the positive odd numbers without multiples of 3, and 1, 2, 4, 5, ..., are the positive integers without multiples of 3. The identities (1.1) and (1.2) are classical and are due to Euler and Jacobi, respectively. The other two identities (1.3) and (1.4) are due to Ramanujan himself. Although Ramanujan gives the exponents explicitly, Watson [107, p. 148] overlooks this and doubts whether Ramanujan was aware of their proofs. Proofs of these two formulas of Ramanujan that use the quintuple product identity will be given at the end of this book in Appendix B.

Let $\tau(n)$ be Ramanujan's tau-function defined by

$$q\prod_{j=1}^{\infty}(1-q^j)^{24} = \sum_{n=1}^{\infty}\tau(n)q^n.$$

At the end of the paper Ramanujan gives the beautiful factorization [91, (101)]:

$$\sum_{n=1}^{\infty}\frac{\tau(n)}{n^s} = \prod_{p \text{ prime}}\frac{1}{1-\tau(p)p^{-s}+p^{11-2s}}$$

and makes the famous conjecture [91, (105)]:

$$|\tau(n)| \le n^{11/2}d(n),$$

where $d(n)$ denotes the number of divisors of n. The conjecture was proved half a century later by Deligne [51], using the advanced methods of algebraic geometry.

In [91, (17)] Ramanujan has given the following identity:

$$\left(\frac{1}{4}\cot\frac{\theta}{2} + \sum_{n=1}^{\infty}\frac{q^n}{1-q^n}\sin n\theta\right)^2 \tag{1.5}$$

$$= \left(\frac{1}{4}\cot\frac{\theta}{2}\right)^2 + \sum_{n=1}^{\infty}\frac{q^n}{(1-q^n)^2}\cos n\theta + \frac{1}{2}\sum_{n=1}^{\infty}\frac{nq^n}{1-q^n}(1-\cos n\theta).$$

He gives a completely algebraic proof of this identity. He also gives the identity [91, (18)]:

$$\left(\frac{1}{8}\cot^2\frac{\theta}{2}+\frac{1}{12}+\sum_{n=1}^{\infty}\frac{nq^n}{1-q^n}(1-\cos n\theta)\right)^2 = \left(\frac{1}{8}\cot^2\frac{\theta}{2}+\frac{1}{12}\right)^2+\frac{1}{12}\sum_{n=1}^{\infty}\frac{n^3q^n}{1-q^n}(5+\cos n\theta), \tag{1.6}$$

and hints that a similar, algebraic proof may be given. The identity (1.6) corresponds to the differential equation satisfied by the Weierstrass elliptic function $\wp(z)$:

$$\wp''(z) = 6\wp^2(z) - g_2/2.$$

Ramanujan obtains many interesting results from the basic identity (1.5), and these are connected with several interesting results in his notebooks. From these identities the basic properties of elliptic functions can be derived in a purely algebraic way without making any use of the Cauchy-Liouville methods of theory of functions. Ramanujan [91, Sec. 7] remarks in a very modest way: *"The elementary proof of these formulæ given in the preceding sections seems to be of some interest in itself"*.

Also, Ramanujan gives the simplest proof of the classical identity:

$$Q^3 - R^2 = 1728q\prod_{j=1}^{\infty}(1-q^j)^{24} \tag{1.7}$$

where

$$Q = 1+240\sum_{j=1}^{\infty}\frac{j^3q^j}{1-q^j} \qquad \text{and} \qquad R = 1-504\sum_{j=1}^{\infty}\frac{j^5q^j}{1-q^j}.$$

In this connection, Ramanujan obtains his important differential equations

$$q\frac{dP}{dq}=\frac{P^2-Q}{12}, \qquad q\frac{dQ}{dq}=\frac{PQ-R}{3} \quad \text{and} \quad q\frac{dR}{dq}=\frac{PR-Q^2}{2},$$

where

$$P = 1-24\sum_{j=1}^{\infty}\frac{jq^j}{1-q^j}.$$

Ramanujan's simple proof of the product formula (1.7) has escaped the attention of scholars. Weil [108, p. 34] and Lang [80, p. 249] base their proofs on the classical identities of Jacobi-Weierstrass theory. Serre, in his excellent monograph [97, p. 95], has repeated Hurwitz's well-known *a priori* proof. All these make use of Cauchy function theory.

A search through older literature reveals that Ramanujan's identity (1.5) was given by Jordan [75, p. 521, eq. (8)] in a somewhat different form. This is obtained after a good deal of elliptic function theory has been developed. Ramanujan, however, makes this identity the foundation of his theory and gives a simple *a priori* proof.

1.2 The generalized Ramanujan identity

The identity (1.5) may be generalized into a form whose proof becomes very simple. Let

$$\rho_1(z) = \frac{1}{2} + \sum_n{}' \frac{z^n}{1-q^n} \quad \text{and} \quad \rho_2(z) = -\frac{1}{12} + \sum_n{}' \frac{q^n z^n}{(1-q^n)^2}, \tag{1.8}$$

where the symbol $\sum_n{}'$ denotes that the summations are over all non-zero integers n. By the ratio test, the series defining ρ_1 converges for $|q| < |z| < 1$, while the series defining ρ_2 converges for $|q| < |z| < |q|^{-1}$. The generalized Ramanujan identity is the following: For any three complex numbers α, β and γ satisfying $\alpha\beta\gamma = 1$ and $|q| < |\alpha|, |\beta|, |\alpha\beta| < 1$, we have

$$\rho_1(\alpha)\rho_1(\beta) - \rho_1(\alpha\beta)(\rho_1(\alpha) + \rho_1(\beta)) = \rho_2(\alpha) + \rho_2(\beta) + \rho_2(\gamma). \tag{1.9}$$

This may be proved as follows. Let the Laurent expansion of the left hand side of (1.9) be

$$\sum_{m,n=-\infty}^{\infty} c_{m,n}\alpha^m\beta^n.$$

Substituting the expansions of $\rho_1(\alpha)$, $\rho_1(\beta)$ and $\rho_1(\alpha\beta)$ in the left hand side of (1.9) we evaluate the coefficients $c_{m,n}$ distinguishing four separate cases.

Case (i): $mn(m-n) \neq 0$. Then using (1.9) we obtain

$$c_{m,n} = \frac{1}{(1-q^m)(1-q^n)} - \frac{1}{(1-q^n)(1-q^{m-n})} - \frac{1}{(1-q^m)(1-q^{n-m})} = 0.$$

Case (ii): If $m \neq 0$ and $n = 0$ then

$$c_{m,0} = \frac{1}{2(1-q^m)} - \frac{1}{2(1-q^m)} - \frac{1}{(1-q^m)(1-q^{-m})} = \frac{q^m}{(1-q^m)^2}.$$

Similarly, if $m = 0$ and $n \neq 0$, then $c_{0,n} = \dfrac{q^n}{(1-q^n)^2}$.

Case (iii): $m = n \neq 0$. Then

$$c_{m,m} = \frac{1}{(1-q^m)^2} - \frac{1}{(1-q^m)} = \frac{q^m}{(1-q^m)^2}.$$

Case (iv): $m = n = 0$. Then $c_{0,0} = 1/4 - 1/2 = -1/4$.

The right hand side of (1.9) consists of three terms besides the constant term, whose coefficients are easily seen to be identical with the corresponding terms of the left hand side that arise from Cases (ii) and (iii). The significant thing we observe is that the bulk of the terms on the left hand side, namely those with $mn(m-n) \neq 0$, cancel out. This completes the proof of the identity (1.9).

We now transform the functions ρ_1 and ρ_2 into appropriate forms which extend the range of their definition. For the function ρ_1, we have:

$$\begin{aligned}\rho_1(z) &= \frac{1}{2} + \sum_{m=1}^{\infty} \frac{z^m}{1-q^m} + \sum_{m=1}^{\infty} \frac{z^{-m}}{1-q^{-m}} \qquad (1.10)\\ &= \frac{1}{2} + \sum_{m=1}^{\infty} \frac{z^m(1-q^m+q^m)}{1-q^m} - \sum_{m=1}^{\infty} \frac{z^{-m}q^m}{1-q^m}\\ &= \frac{1}{2} + \frac{z}{1-z} + \sum_{m=1}^{\infty} \frac{q^m}{1-q^m}(z^m - z^{-m}).\end{aligned}$$

This can be expanded as a double series:

$$\rho_1(z) = \frac{1+z}{2(1-z)} + \sum_{m=1}^{\infty}\sum_{n=1}^{\infty}(z^m q^{mn} - z^{-m}q^{mn}).$$

We interchange the order of summation in the above and obtain

$$\rho_1(z) = \frac{1+z}{2(1-z)} + \sum_{n=1}^{\infty}\left(\frac{zq^n}{1-zq^n} - \frac{z^{-1}q^n}{1-z^{-1}q^n}\right). \qquad (1.11)$$

This shows that ρ_1 is analytic in $0 < |z| < \infty$ except at $z = q^r$, $r = 0, \pm1, \pm2, \ldots$, where there are simple poles. There is an essential singularity at $z = 0$. From (1.11) we may deduce the symmetric property

$$\rho_1(z) + \rho_1(z^{-1}) = 0,$$

and the quasi-periodic property

$$\rho_1(z) - \rho_1(qz) = -1.$$

For the function ρ_2, we have:

$$\begin{aligned}\rho_2(z) &= -\frac{1}{12} + \sum_{m=1}^{\infty} \frac{q^m}{(1-q^m)^2}(z^m + z^{-m}) \\ &= -\frac{1}{12} + \sum_{m=1}^{\infty}\sum_{n=1}^{\infty} nq^{mn}(z^m + z^{-m}) \\ &= -\frac{1}{12} + \sum_{n=1}^{\infty}\left(\frac{nzq^n}{1-zq^n} + \frac{nz^{-1}q^n}{1-z^{-1}q^n}\right).\end{aligned} \tag{1.12}$$

Therefore, ρ_2 has simple poles at the points $z = q^r$, $r = \pm 1, \pm 2, \ldots$, and is otherwise analytic in $0 < |z| < \infty$, with an essential singularity at $z = 0$. The symmetry property

$$\rho_2(z) = \rho_2(z^{-1})$$

is an immediate consequence of (1.12). Furthermore, using (1.11) and (1.12) we find that the quasi-periodic property for ρ_2 is given by

$$\begin{aligned}\rho_2(z) - \rho_2(qz) &= \sum_{n=1}^{\infty}\left(\frac{nzq^n}{1-zq^n} + \frac{nz^{-1}q^n}{1-z^{-1}q^n} - \frac{nzq^{n+1}}{1-zq^{n+1}} - \frac{nz^{-1}q^{n-1}}{1-z^{-1}q^{n-1}}\right) \\ &= \frac{1}{1-z} + \sum_{n=1}^{\infty}\left(\frac{zq^n}{1-zq^n} - \frac{z^{-1}q^n}{1-z^{-1}q^n}\right) \\ &= \frac{1}{2} + \rho_1(z).\end{aligned}$$

The generalized Ramanujan identity (1.9) can be extended to all values of α, β and γ by using the symmetric and quasi-periodic properties of ρ_1 and ρ_2. The identity may be written in the symmetric form

$$\rho_1(\alpha)\rho_1(\beta) + \rho_1(\beta)\rho_1(\gamma) + \rho_1(\gamma)\rho_1(\alpha) = \rho_2(\alpha) + \rho_2(\beta) + \rho_2(\gamma), \tag{1.13}$$

which holds for all complex numbers α, β and γ that satisfy $\alpha\beta\gamma = 1$.

For future reference we note the limiting case of (1.9) in which $\beta \to 1$. First, from (1.10),

$$\lim_{\beta\to 1}(1-\beta)\rho_1(\beta) = 1,$$

and so

$$\lim_{\beta\to 1}\rho_1(\beta)(\rho_1(\alpha)-\rho_1(\alpha\beta)) = \lim_{\beta\to 1}(1-\beta)\rho_1(\beta)\times\lim_{\beta\to 1}\frac{\rho_1(\alpha)-\rho_1(\alpha\beta)}{1-\beta} = \alpha\rho_1'(\alpha).$$

Hence, making $\beta \to 1$ in (1.9), we obtain

$$\alpha\rho_1'(\alpha) - \rho_1^2(\alpha) = 2\rho_2(\alpha) + \rho_2(1). \tag{1.14}$$

Let ϕ_1 and ϕ_2 be the functions defined by

$$\phi_1(\theta) = \frac{1}{2i}\rho_1(e^{i\theta}) \qquad \text{and} \qquad \phi_2(\theta) = \frac{1}{2}\rho_2(e^{i\theta}). \tag{1.15}$$

From (1.10) and (1.12), we immediately deduce the Fourier expansions

$$\phi_1(\theta) = \frac{1}{4}\cot\frac{\theta}{2} + \sum_{n=1}^{\infty} \frac{q^n}{1-q^n}\sin n\theta \tag{1.16}$$

and

$$\phi_2(\theta) = -\frac{1}{24} + \sum_{n=1}^{\infty} \frac{q^n}{(1-q^n)^2}\cos n\theta. \tag{1.17}$$

We put $q = \exp(2\pi i\tau)$, where $\mathrm{Im}(\tau) > 0$. The following properties of ϕ_1 and ϕ_2 follow immediately from the corresponding properties of ρ_1 and ρ_2:

$$\phi_1(-\theta) = -\phi_1(\theta), \qquad \phi_2(-\theta) = \phi_2(\theta), \tag{1.18}$$

$$\phi_1(\theta + 2\pi) = \phi_1(\theta), \qquad \phi_2(\theta + 2\pi) = \phi_2(\theta), \tag{1.19}$$

$$\phi_1(\theta + 2\pi\tau) = \phi_1(\theta) - \frac{i}{2}, \qquad \phi_2(\theta + 2\pi\tau) = \phi_2(\theta) - i\phi_1(\theta) - \frac{1}{4}. \tag{1.20}$$

The function $\phi_1(\theta)$ has simple poles at each point $\theta = 2\pi m + 2\pi n\tau$ for all integers m and n and no other singularities, and the residue at each pole is $1/2$.

The generalized Ramanujan identity (1.13) assumes the form

$$\phi_1(a)\phi_1(b) + \phi_1(b)\phi_1(c) + \phi_1(c)\phi_1(a) = -\frac{1}{2}(\phi_2(a) + \phi_2(b) + \phi_2(c)), \tag{1.21}$$

provided $a + b + c = 0$. The identity (1.14) is equivalent to

$$\phi_1^2(a) + \frac{1}{2}\phi_1'(a) = \phi_2(a) + \frac{1}{2}\phi_2(0). \tag{1.22}$$

This can be seen to be equivalent to Ramanujan's identity (1.5), on using the Fourier expansions for ϕ_1 and ϕ_2 in (1.16) and (1.17).

The new functions ϕ_1 and ϕ_2 and the generalized Ramanujan identity (1.21) form the foundation for the development of elliptic functions. Its form indicates that (1.21) is of relevance in obtaining some quadratic transformation formulas; in fact Ramanujan [91, (20)] obtained the sum of four squares formula from the sum of two squares formula using (1.5) and deduced several other interesting results.

1.3 The Weierstrass elliptic function

We will show how the generalized Ramanujan identity in the form (1.21) can be used to derive the addition formula and differential equation for the Weierstrass elliptic function. We begin with the addition formula. We substitute b and c in place of a in (1.22) and add to obtain

$$\phi_1^2(a)+\phi_1^2(b)+\phi_1^2(c)+\frac{1}{2}(\phi_1'(a)+\phi_1'(b)+\phi_1'(c)) = \phi_2(a)+\phi_2(b)+\phi_2(c)+\frac{3}{2}\phi_2(0). \tag{1.23}$$

Now suppose $a+b+c=0$. We multiply (1.21) by 2 and add to (1.23) to get

$$(\phi_1(a)+\phi_1(b)+\phi_1(c))^2+\frac{1}{2}(\phi_1'(a)+\phi_1'(b)+\phi_1'(c)) = \frac{3}{2}\phi_2(0). \tag{1.24}$$

We apply the differential operator $\partial/\partial a - \partial/\partial b$ to obtain

$$2(\phi_1(a)+\phi_1(b)+\phi_1(c))(\phi_1'(a)-\phi_1'(b))+\frac{1}{2}(\phi_1''(a)-\phi_1''(b)) = 0, \tag{1.25}$$

or

$$\phi_1(a)+\phi_1(b)+\phi_1(c) = -\frac{\phi_1''(a)-\phi_1''(b)}{4(\phi_1'(a)-\phi_1'(b))}. \tag{1.26}$$

Combining equations (1.24) and (1.26) gives

$$\frac{1}{16}\left(\frac{\phi_1''(a)-\phi_1''(b)}{\phi_1'(a)-\phi_1'(b)}\right)^2 = \frac{3}{2}\phi_2(0)-\frac{1}{2}(\phi_1'(a)+\phi_1'(b)+\phi_1'(c)). \tag{1.27}$$

We define[1] the Weierstrass elliptic function $\wp$ by

$$\wp(a) = 2(\phi_2(0)-\phi_1'(a)). \tag{1.28}$$

Then (1.27) becomes

$$\wp(a+b) = \frac{1}{4}\left(\frac{\wp'(a)-\wp'(b)}{\wp(a)-\wp(b)}\right)^2 - \wp(a)-\wp(b). \tag{1.29}$$

This is the addition theorem for the Weierstrass elliptic function.

The Abelian form of the addition theorem can be derived similarly. Apply $\partial^2/\partial a\partial b$ to (1.25) to obtain

$$\begin{vmatrix} 1 & 1 & 1 \\ \phi_1'(a) & \phi_1'(b) & \phi_1'(c) \\ \phi_1''(a) & \phi_1''(b) & \phi_1''(c) \end{vmatrix} = 0, \qquad \text{where} \quad a+b+c=0.$$

[1]The standard definition of the Weierstrass elliptic function will be given in the notes at the end of Chapter 1, and its equivalence with the definition in (1.28) will be established there.

This, expressed in terms of the Weierstrass function, assumes the form

$$\begin{vmatrix} 1 & 1 & 1 \\ \wp(a) & \wp(b) & \wp(c) \\ \wp'(a) & \wp'(b) & \wp'(c) \end{vmatrix} = 0, \qquad \text{where} \quad a+b+c=0.$$

In order to obtain the differential equation for the Weierstrass function, we will first need to expand $\phi_1(\theta)$ and $\phi_2(\theta)$ in powers of θ. We will use the standard notation for the Bernoulli numbers B_n given by

$$\frac{\theta}{e^\theta - 1} = \sum_{n=0}^{\infty} \frac{B_n}{n!}\theta^n. \tag{1.30}$$

The first few even Bernoulli numbers are given by

$$\begin{gathered} B_0 = 1, \;\; B_2 = \frac{1}{6}, \;\; B_4 = -\frac{1}{30}, \;\; B_6 = \frac{1}{42}, \\ B_8 = -\frac{1}{30}, \;\; B_{10} = \frac{5}{66}, \;\; B_{12} = -\frac{691}{2730}. \end{gathered} \tag{1.31}$$

The odd Bernoulli numbers are given by

$$B_1 = -\frac{1}{2}, \quad B_3 = B_5 = B_7 = \cdots = 0.$$

The series expansion of the cotangent function is given by

$$\frac{1}{2}\cot\frac{\theta}{2} = \sum_{n=0}^{\infty} \frac{(-1)^n B_{2n}}{(2n)!}\theta^{2n-1}.$$

The expansion of $\phi_1(\theta)$ in powers of θ may now be obtained by expanding the Fourier expansion (1.16) in powers of θ. We find that

$$\begin{aligned} \phi_1(\theta) &= \frac{1}{4}\cot\frac{\theta}{2} + \sum_{j=1}^{\infty} \frac{q^j}{1-q^j}\sin j\theta \\ &= \frac{1}{2\theta} + \sum_{n=1}^{\infty} \frac{(-1)^{n-1}}{(2n-1)!}\left\{-\frac{B_{2n}}{4n} + \sum_{j=1}^{\infty}\frac{j^{2n-1}q^j}{1-q^j}\right\}\theta^{2n-1}. \end{aligned} \tag{1.32}$$

The initial terms in the expansion are given by

$$\begin{aligned} \phi_1(\theta) &= \frac{1}{2\theta} + \left\{-\frac{1}{24} + \sum_{j=1}^{\infty}\frac{jq^j}{1-q^j}\right\}\theta \\ &\quad - \left\{\frac{1}{240} + \sum_{j=1}^{\infty}\frac{j^3q^j}{1-q^j}\right\}\frac{\theta^3}{3!} + \left\{-\frac{1}{504} + \sum_{j=1}^{\infty}\frac{j^5q^j}{1-q^j}\right\}\frac{\theta^5}{5!} - \cdots \\ &= \frac{1}{2\theta} - \left(\frac{P}{24}\right)\theta - \left(\frac{Q}{240}\right)\left(\frac{\theta^3}{3!}\right) - \left(\frac{R}{504}\right)\left(\frac{\theta^5}{5!}\right) - \cdots, \end{aligned} \tag{1.33}$$

where P, Q and R are Ramanujan's Eisenstein series defined by

$$P = 1 - 24\sum_{j=1}^{\infty}\frac{jq^j}{1-q^j}, \quad Q = 1 + 240\sum_{j=1}^{\infty}\frac{j^3q^j}{1-q^j}, \quad R = 1 - 504\sum_{j=1}^{\infty}\frac{j^5q^j}{1-q^j}. \tag{1.34}$$

We expand (1.17) in powers of θ, to get

$$\begin{aligned}\phi_2(\theta) &= -\frac{1}{24} + \sum_{j=1}^{\infty}\frac{q^j}{(1-q^j)^2}\cos j\theta \qquad (1.35)\\ &= \left\{-\frac{1}{24} + \sum_{j=1}^{\infty}\frac{q^j}{(1-q^j)^2}\right\} + \sum_{n=1}^{\infty}\frac{(-1)^n}{(2n)!}\left\{\sum_{j=1}^{\infty}\frac{j^{2n}q^j}{(1-q^j)^2}\right\}\theta^{2n}.\end{aligned}$$

Now

$$\begin{aligned}\phi_2(0) &= -\frac{1}{24} + \sum_{j=1}^{\infty}\frac{q^j}{(1-q^j)^2} = -\frac{1}{24} + \sum_{j=1}^{\infty}\sum_{k=1}^{\infty}kq^{jk} = -\frac{1}{24} + \sum_{k=1}^{\infty}\frac{kq^k}{1-q^k}\\ &= -\frac{P}{24}, \qquad (1.36)\end{aligned}$$

and

$$\sum_{j=1}^{\infty}\frac{j^{2n}q^j}{(1-q^j)^2} = \sum_{j=1}^{\infty}\sum_{k=1}^{\infty}j^{2n}kq^{jk} = \phi_{2n,1},$$

where $\phi_{m,n}$ is the series defined by Ramanujan by

$$\phi_{m,n} = \sum_{j=1}^{\infty}\sum_{k=1}^{\infty}j^m k^n q^{jk} \quad (= \phi_{n,m}).$$

Therefore, the first two terms in the expansion (1.35) are given by

$$\phi_2(\theta) = -\frac{P}{24} - \phi_{1,2}\frac{\theta^2}{2!} + \cdots. \tag{1.37}$$

The first few terms in the expansion of the Weierstrass function may be determined using (1.28), (1.33) and (1.36), and they are given by:

$$\begin{aligned}\wp(\theta) &= 2\left(\phi_2(0) - \phi_1'(\theta)\right) \qquad (1.38)\\ &= \frac{1}{\theta^2} - 2\sum_{n=1}^{\infty}\frac{(-1)^n}{(2n)!}\left\{-\frac{B_{2n+2}}{4(n+1)} + \sum_{j=1}^{\infty}\frac{j^{2n+1}q^j}{1-q^j}\right\}\theta^{2n}\\ &= \frac{1}{\theta^2} + \frac{Q}{240}\theta^2 + \frac{R}{12\times 504}\theta^4 + \cdots.\end{aligned}$$

Now we will obtain the differential equation satisfied by the Weierstrass elliptic function. Write the generalized Ramanujan identity (1.21) in the form

$$(\phi_1(a)+\phi_1(b))\phi_1(a+b)-\phi_1(a)\phi_1(b)=\frac{1}{2}(\phi_2(a)+\phi_2(b)+\phi_2(a+b)).$$

Expand both sides in powers of b and compare the coefficients of b^2 to obtain

$$2\phi_1(a)\phi_1''(a)-\frac{P}{6}\phi_1'(a)+\frac{1}{3}\phi_1'''(a)=\phi_2''(a)-\phi_{1,2}.$$

Next, differentiate (1.22) twice to get

$$2\phi_1(a)\phi_1''(a)+2(\phi_1'(a))^2+\frac{1}{2}\phi_1'''(a)=\phi_2''(a).$$

Eliminating $\phi_2''(a)$ from the last two equations gives

$$4(\phi_1'(a))^2+\frac{P}{3}\phi_1'(a)+\frac{1}{3}\phi_1'''(a)=2\phi_{1,2},$$

which we write as

$$\left(2\phi_1'(a)+\frac{P}{12}\right)^2+\frac{1}{3}\phi_1'''(a)=2\phi_{1,2}+\frac{P^2}{144}.$$

Using the definition of the Weierstrass function (1.28), this is equivalent to

$$\wp^2(a)-\frac{1}{6}\wp''(a)=2\phi_{1,2}+\frac{P^2}{144}. \tag{1.39}$$

We use (1.38) to expand both sides of (1.39), and then equate the constant terms, to get

$$\frac{Q}{144}=2\phi_{1,2}+\frac{P^2}{144}.$$

Consequently,

$$\wp^2(a)-\frac{1}{6}\wp''(a)=\frac{Q}{144}. \tag{1.40}$$

This is equivalent to (1.6). Multiply both sides by $12\wp'(a)$ and integrate, to obtain

$$(\wp'(a))^2=4\wp^3(a)-\frac{Q}{12}\wp(a)+k, \tag{1.41}$$

for some constant k. Once again we use (1.38) to expand (1.41) in powers of a, and equate constant terms. This gives

$$k=-\frac{R}{216}.$$

Therefore

$$(\wp'(a))^2=4\wp^3(a)-\frac{Q}{12}\wp(a)-\frac{R}{216}. \tag{1.42}$$

This is the differential equation for the Weierstrass elliptic function. Putting $g_2=Q/12$ and $g_3=R/216$ we obtain the differential equation in its standard form:

$$(\wp'(a))^2=4\wp^3(a)-g_2\wp(a)-g_3. \tag{1.43}$$

1.4 Notes

In several instances, reference has been made to results in Ramanujan's paper [91]. The reader is referred to the commentary by B. C. Berndt [94, pp. 365–368] for a summary of the influence of Ramanujan's paper and the activity it has generated.

Ramanujan's identities (1.3) and (1.4) are on p. 266 of his first notebook [92], in Chapter 17, Entries 8 (ix) and (x) in his second notebook [92], and in his paper [91, eq. (65)].

Ramanujan's identity (1.5) can be expressed in terms of theta functions, as follows. It is easy to check that the theta function

$$u(\theta,t) = -i\sum_{j=-\infty}^{\infty}(-1)^j q^{(2j+1)^2/8}e^{i(2j+1)\theta/2}, \quad \text{where} \quad q=e^{-t}, \tag{1.44}$$

is a solution of the heat equation

$$2\frac{\partial u}{\partial t} = \frac{\partial^2 u}{\partial \theta^2}.$$

Therefore, $\log u$ satisfies the non-linear partial differential equation

$$\left(\frac{\partial \log u}{\partial \theta}\right)^2 = 2\frac{\partial \log u}{\partial t} - \frac{\partial^2 \log u}{\partial \theta^2}. \tag{1.45}$$

In the next chapter we will prove Jacobi's triple product identity (2.36). This identity, which we assume for now, implies that the series defined by (1.44) has the factorization given by

$$u(\theta,t) = 2q^{1/8}\sin\frac{\theta}{2}\prod_{j=1}^{\infty}(1-e^{i\theta}q^j)(1-e^{-i\theta}q^j)(1-q^j). \tag{1.46}$$

In 1951, B. van der Pol [101] observed that if (1.46) is substituted into (1.45), the result is Ramanujan's identity (1.5).

Ramanujan's own proof of (1.5) in [91] has been described by G. H. Hardy [62, p. 133] as *"a very characteristic specimen of Ramanujan's work"*. Hardy liked Ramanujan's proof so much that he reproduced it in his lectures [62, pp. 133–135] as well as in his book on number theory with E. M. Wright [63, pp. 312–314].

Proofs of Ramanujan's identity (1.5) and an extension, by contour integration, have been given by V. Ramamani [87, pp. 82–92].

The definition of the Weierstrass elliptic function given by (1.28) is not the usual one. The standard definition of the Weierstrass elliptic function with periods 2π and $2\pi\tau$, where $\operatorname{Im}(\tau) > 0$, is given by

$$\wp(\theta) = \frac{1}{\theta^2} + \sum_{m,n}{}'\left(\frac{1}{(\theta-2\pi n-2\pi\tau m)^2} - \frac{1}{(2\pi n+2\pi\tau m)^2}\right), \tag{1.47}$$

where the symbol $\sum_{m,n}'$ indicates that the summation is over all integers m and n with the term corresponding to $(m,n)=(0,0)$ omitted. The basic properties of the series defined by (1.47) are the following: the series converges absolutely and uniformly on compact sets that do not contain any of the points $2\pi n+2\pi m\tau$, $m,n\in\mathbb{Z}$; the function $\wp(\theta)$ is analytic except at the points $\theta=2\pi n+2\pi m\tau$, $m,n\in\mathbb{Z}$, where there are poles of order 2; and $\wp(\theta)$ is a doubly periodic function of θ with periods 2π and $2\pi\tau$. For proofs of these standard facts, see, for example, the book by Whittaker and Watson [111, pp. 434–435].

We will now show that the equations (1.28) and (1.47) are equivalent. Recall the standard results [3, pp. 11, 12] or [36, pp. 218–222]:

$$\sum_{n=-\infty}^{\infty}\frac{1}{(\theta-2\pi n)^2}=\frac{1}{4\sin^2\frac{\theta}{2}}\qquad\text{and}\qquad\sum_{n=1}^{\infty}\frac{1}{n^2}=\frac{\pi^2}{6}.$$

Then, considering the terms $m=0$ and $m\neq 0$ separately, we have from (1.47) that

$$\begin{aligned}
\wp(\theta)&=\frac{1}{\theta^2}+\sum_{\substack{n=-\infty\\ n\neq 0}}^{\infty}\left(\frac{1}{(\theta-2\pi n)^2}-\frac{1}{(2\pi n)^2}\right)\\
&\quad+\sum_{\substack{m=-\infty\\ m\neq 0}}^{\infty}\sum_{n=-\infty}^{\infty}\left(\frac{1}{(\theta-2\pi n-2\pi\tau m)^2}-\frac{1}{(2\pi n+2\pi\tau m)^2}\right)\\
&=\sum_{n=-\infty}^{\infty}\frac{1}{(\theta-2\pi n)^2}-\frac{1}{2\pi^2}\sum_{n=1}^{\infty}\frac{1}{n^2}\\
&\quad+\sum_{\substack{m=-\infty\\ m\neq 0}}^{\infty}\sum_{n=-\infty}^{\infty}\left(\frac{1}{(\theta-2\pi n-2\pi\tau m)^2}-\frac{1}{(2\pi n+2\pi\tau m)^2}\right)\\
&=\frac{1}{4\sin^2\frac{\theta}{2}}-\frac{1}{12}+\sum_{\substack{m=-\infty\\ m\neq 0}}^{\infty}\left(\frac{1}{4\sin^2(\frac{\theta}{2}-\pi\tau m)}-\frac{1}{4\sin^2\pi\tau m}\right)\\
&=-\frac{1}{12}-\frac{1}{2}\sum_{m=1}^{\infty}\frac{1}{\sin^2\pi\tau m}+\frac{1}{4}\sum_{m=-\infty}^{\infty}\frac{1}{\sin^2(\frac{\theta}{2}+\pi\tau m)}.
\end{aligned}$$

Since $q = e^{2\pi i\tau}$, this is equivalent to

$$\begin{aligned}\wp(\theta) &= -\frac{1}{12} + 2\sum_{m=1}^{\infty} \frac{1}{(q^{m/2} - q^{-m/2})^2} \\ &\quad - \sum_{m=-\infty}^{\infty} \frac{1}{(e^{i\theta/2}q^{m/2} - e^{-i\theta/2}q^{-m/2})^2} \\ &= -\frac{1}{12} + 2\sum_{m=1}^{\infty} \frac{q^m}{(1-q^m)^2} - \sum_{m=-\infty}^{\infty} \frac{e^{i\theta}q^m}{(1-e^{i\theta}q^m)^2}.\end{aligned}$$

This can be manipulated further, using (1.36), to give

$$\begin{aligned}\wp(\theta) &= 2\phi_2(0) - \frac{e^{i\theta}}{(1-e^{i\theta})^2} - \sum_{m=1}^{\infty}\left(\frac{e^{i\theta}q^m}{(1-e^{i\theta}q^m)^2} + \frac{e^{-i\theta}q^m}{(1-e^{-i\theta}q^m)^2}\right) \\ &= 2\phi_2(0) + i\frac{d}{d\theta}\left\{\frac{e^{i\theta}}{1-e^{i\theta}} + \sum_{m=1}^{\infty}\left(\frac{e^{i\theta}q^m}{1-e^{i\theta}q^m} - \frac{e^{-i\theta}q^m}{1-e^{-i\theta}q^m}\right)\right\}.\end{aligned}$$

Finally, applying (1.11) and (1.15) we get

$$\wp(\theta) = 2\phi_2(0) + i\frac{d}{d\theta}\left\{\rho_1(e^{i\theta}) - \frac{1}{2}\right\} = 2\left(\phi_2(0) - \phi_1'(\theta)\right).$$

Thus, (1.28) follows from (1.47). The steps are reversible, and so (1.28) and (1.47) may be regarded as being equivalent.

Chapter 2

The Differential Equations of P, Q and R

2.1 Ramanujan's differential equations

Ramanujan proved that the functions P, Q and R in (1.34) satisfy the system of differential equations

$$q\frac{dP}{dq} = \frac{P^2 - Q}{12}, \qquad q\frac{dQ}{dq} = \frac{PQ - R}{3} \quad \text{and} \quad q\frac{dR}{dq} = \frac{PR - Q^2}{2}.$$

These were used by him to derive several classical identities and some new ones, too. For example, let

$$\phi_{m,n} = \sum_{j=1}^{\infty}\sum_{k=1}^{\infty} j^m k^n q^{jk} \quad (= \phi_{n,m}) \tag{2.1}$$

and

$$S_r = -\frac{B_{r+1}}{2(r+1)} + \phi_{0,r}, \tag{2.2}$$

where m, n and r are non-negative integers, and B_n denotes the nth Bernoulli number. Ramanujan has proved the following:

Theorem 2.1.

(i) *If $r \geq 1$, then S_{2r+1} is a polynomial in Q and R over the field of rational numbers. This is a classical result, e.g., see the lectures of K. Weierstrass written up and published by H. A. Schwarz [96, pp. 10, 11].*

(ii) *If $m + n$ is odd, then $\phi_{m,n}$ is a polynomial in P, Q and R over the field of rational numbers. This is a new result, discovered by Ramanujan.*

We will now give Ramanujan's proof of Theorem 2.1, as well as his proof of the differential equations for P, Q and R.

With the notation defined above, the series expansions for ϕ_1, ϕ_2 and $\wp$ given by (1.32), (1.35) and (1.38) can be written as

$$\phi_1(\theta) = \frac{1}{2\theta} + \sum_{n=1}^{\infty} \frac{(-1)^{n-1}}{(2n-1)!} S_{2n-1}\theta^{2n-1}, \tag{2.3}$$

$$\phi_2(\theta) = -\frac{1}{24} + \sum_{n=0}^{\infty} \frac{(-1)^n}{(2n)!} \phi_{1,2n}\theta^{2n}, \tag{2.4}$$

and

$$\wp(\theta) = \frac{1}{\theta^2} + 2\sum_{n=1}^{\infty} \frac{(-1)^{n+1}}{(2n)!} S_{2n+1}\theta^{2n}. \tag{2.5}$$

Also, we have

$$\begin{cases} S_1 = & -\dfrac{1}{24} + \phi_{0,1} = & -\dfrac{P}{24}, \\ S_3 = & \dfrac{1}{240} + \phi_{0,3} = & \dfrac{Q}{240}, \\ S_5 = & -\dfrac{1}{504} + \phi_{0,5} = & -\dfrac{R}{504}. \end{cases} \tag{2.6}$$

We recall the differential equation (1.40):

$$\wp''(\theta) = 6\wp^2(\theta) - \frac{Q}{24}.$$

Expanding both sides in powers of θ using (2.5) and comparing coefficients of θ^{2n} on both sides, we obtain

$$S_{2n+3} = \frac{12(n+1)(2n+1)}{(n-1)(2n+5)} \sum_{j=1}^{n-1} \binom{2n}{2j} S_{2j+1}S_{2n+1-2j}, \qquad n = 2, 3, \ldots. \tag{2.7}$$

For example, taking $n = 2$ gives

$$S_7 = 120S_3^2. \tag{2.8}$$

By induction using (2.7), we can express S_9, S_{11}, S_{13}, ..., as polynomials in S_3 and S_5 (equivalently, as polynomials in Q and R), with rational coefficients. In general we have

$$S_{2n-1} = -\frac{B_{2n}}{4n} + \phi_{0,2n-1} = \sum K_{i,j}Q^iR^j \tag{2.9}$$

where $K_{i,j}$ is a rational number and i and j are non-negative integers satisfying the condition $2i + 3j = n$, and $n \geq 2$. This proves statement (i) in Theorem 2.1.

We now utilize Ramanujan's identity (1.5), which we write in an equivalent form using (1.22), (1.36) and (2.6), as

$$\phi_1^2(\theta) + \frac{1}{2}\phi_1'(\theta) = \phi_2(\theta) + \frac{S_1}{2}.$$

Expanding both sides in powers of θ using (2.3) and (2.4) and comparing coefficients of θ^{2n}, we obtain

$$\phi_{1,2n} = \frac{2n+3}{2(2n+1)}S_{2n+1} - \sum_{j=1}^{n} \binom{2n}{2j-1} S_{2j-1}S_{2n+1-2j}, \qquad n = 1, 2, \ldots. \tag{2.10}$$

It follows that $\phi_{1,2n}$ is a polynomial in S_1, S_3, ..., S_{2n+1}, with rational coefficients. Since S_7, S_9, ..., are polynomials in S_3 and S_5, it follows that $\phi_{1,n}$ is a polynomial in P, Q and R, with rational coefficients.

Taking $n = 1$, 2 and 3 in (2.10), we obtain

$$\begin{aligned}
\phi_{1,2} &= \frac{5}{6}S_3 - 2S_1^2, \\
\phi_{1,4} &= \frac{7}{10}S_5 - 8S_1S_3, \\
\phi_{1,6} &= \frac{9}{14}S_7 - 12S_1S_5 - 20S_3^2 \;=\; \frac{400}{7}S_3^2 - 12S_1S_5,
\end{aligned}$$

where (2.8) has been used in the last step. By (2.6), these are equivalent to

$$288\phi_{1,2} = Q - P^2, \quad 720\phi_{1,4} = PQ - R, \quad 1008\phi_{1,6} = Q^2 - PR. \tag{2.11}$$

By induction it may be shown that

$$\phi_{1,2n} = \sum K_{i,j,\ell}P^iQ^jR^\ell,$$

where $K_{i,j,\ell}$ is a rational number, i, j and ℓ are non-negative integers that satisfy $i + 2j + 3\ell = n + 1$, and $n \geq 1$. Furthermore, if $n \geq 2$, then $i \leq 1$.

We now observe from the definition (2.1) that

$$q\frac{d}{dq}\phi_{m,n} = \phi_{m+1,n+1}. \tag{2.12}$$

Therefore, the identities in (2.11) may be rewritten in the form

$$288\,q\frac{d}{dq}\phi_{0,1} = Q - P^2, \quad 720\,q\frac{d}{dq}\phi_{0,3} = PQ - R, \quad 1008\,q\frac{d}{dq}\phi_{0,5} = Q^2 - PR,$$

and so, on applying (2.6) we deduce that

$$q\frac{dP}{dq} = \frac{P^2 - Q}{12}, \qquad q\frac{dQ}{dq} = \frac{PQ - R}{3}, \qquad q\frac{dR}{dq} = \frac{PR - Q^2}{2}. \tag{2.13}$$

These are called Ramanujan's differential equations.

The discriminant of the polynomial equation $a(t-r_1)(t-r_2)\cdots(t-r_n) = 0$ is defined to be [52, p. 152]:

$$a^{2n-2} \prod_{1\le i<j\le n} (r_i - r_j)^2.$$

The discriminant of the cubic equation $t^3 - pt - q = 0$ is known to be $4p^3 - 27q^2$; see [52, p. 153]. Therefore, the discriminant of the Weierstrassian cubic (see (1.42))

$$4t^3 - \frac{Q}{12}t - \frac{R}{216} = 0$$

is a constant multiple of $Q^3 - R^2$. An immediate consequence of Ramanujan's differential equations is

$$\begin{aligned} q\frac{d}{dq}\log(Q^3 - R^2) &= \frac{1}{Q^3 - R^2}\left(3Q^2\, q\frac{dQ}{dq} - 2R\, q\frac{dR}{dq}\right) \\ &= \frac{1}{Q^3 - R^2}\left(Q^2(PQ - R) - R(PR - Q^2)\right) \\ &= P \\ &= 1 - 24\sum_{n=1}^{\infty} \frac{nq^n}{1 - q^n}. \end{aligned} \tag{2.14}$$

Integrating, we obtain

$$\log(Q^3 - R^2) = c + \log q + 24\sum_{n=1}^{\infty}\log(1 - q^n)$$

for some constant c, and so

$$Q^3 - R^2 = b\,q\prod_{n=1}^{\infty}(1 - q^n)^{24}$$

for some constant b. Comparing the coefficients of q on both sides gives

$$b = 3(240) + 2(504) = 1728.$$

This proves the factorization formula of the discriminant:

$$Q^3 - R^2 = 1728q\prod_{n=1}^{\infty}(1 - q^n)^{24}. \tag{2.15}$$

Most proofs of this celebrated formula use a good deal of the development of the theory of elliptic functions or the theory of elliptic modular functions, both of which rely on Cauchy's function theory. However Ramanujan's algebraic proof given above, which relies on the differential equations (2.13), is the simplest.

Ramanujan's tau-function $\tau(n)$ is defined by the formula

$$q\prod_{j=1}^{\infty}(1-q^j)^{24}=\sum_{n=1}^{\infty}\tau(n)q^n. \tag{2.16}$$

Ramanujan was the first mathematician to observe some interesting congruence properties of $\tau(n)$, and also of the partition function $p(n)$ that we shall meet in Sec. 3.3. For example, he proved that

$$\tau(n)\equiv\sigma_{11}(n)\pmod{691}, \tag{2.17}$$

where

$$\sigma_r(n)=\sum_{d|n}d^r.$$

This may be proved using the results we have found so far. We let $n=6$ in (2.9) and observe that there will only be two terms on the right hand side, to get

$$-\frac{B_{12}}{24}+\sum_{j=1}^{\infty}\sum_{k=1}^{\infty}j^{11}q^{jk}=c_1Q^3+c_2R^2, \tag{2.18}$$

for some constants c_1 and c_2. The value of B_{12} was given in (1.31), and the values of c_1 and c_2 can be determined by equating the constant term and the coefficient of q on both sides of (2.18). The resulting formula may be expressed in the form

$$691+65520\sum_{n=1}^{\infty}\sigma_{11}(n)q^n=441Q^3+250R^2. \tag{2.19}$$

Next, by (2.15) and (2.16), we have

$$1728\sum_{n=1}^{\infty}\tau(n)q^n=Q^3-R^2. \tag{2.20}$$

If we multiply (2.19) by 12, multiply (2.20) by 455, then subtract one from the other, we obtain

$$8292\;+\;2^6\times3^3\times5\times7\times13\sum_{n=1}^{\infty}\left(\sigma_{11}(n)-\tau(n)\right)q^n=691(7Q^3+5R^2).$$

As 691 is prime, this proves Ramanujan's congruence (2.17).

We will now prove Ramanujan's claim (ii) in Theorem 2.1. Suppose $m < n$ and that $m+n$ is odd. Clearly,

$$\phi_{m,n} = \left(q\frac{d}{dq}\right)^m \phi_{0,n-m}.$$

Since $\phi_{0,n-m}$, $q\frac{dP}{dq}$, $q\frac{dQ}{dq}$ and $q\frac{dR}{dq}$ are all polynomials in P, Q and R, it follows on using the differential equations (2.13) that $\phi_{m,n}$ is a polynomial in P, Q and R. This proves Ramanujan's result (ii) in Theorem 2.1.

2.2 Ramanujan's ${}_1\psi_1$ summation formula

Functions closely allied to the elliptic functions are the so-called theta functions. They occur in Bernoulli's work, and significant results concerning them are found in the works of Euler and Gauss. But Jacobi was the first mathematician to study them exhaustively and establish their connections with elliptic functions. In fact he developed them purely algebraically and based the theory of elliptic functions on their detailed investigation. One of the most significant results in this connection is the Jacobi triple product identity. In one form, this is

$$\prod_{n=1}^{\infty}(1+zq^{2n-1})(1+z^{-1}q^{2n-1})(1-q^{2n}) = \sum_{n=-\infty}^{\infty} q^{n^2} z^n, \tag{2.21}$$

where $|q| < 1$ and $z \neq 0$. We will consider the primary variable to be z, and regard q as a parameter. Consider the factors in the infinite product that involve z, and let f be the function defined by

$$f(z) = \prod_{n=1}^{\infty}(1+zq^{2n-1})(1+z^{-1}q^{2n-1}). \tag{2.22}$$

Since the product converges absolutely and uniformly on compact subsets of the region $0 < |z| < \infty$, it has a Laurent expansion of the form

$$f(z) = \sum_{n=-\infty}^{\infty} c_n z^n$$

in the annulus $0 < |z| < \infty$. One can easily show that

$$f(z) = qzf(zq^2).$$

Expanding both sides in powers of z and equating coefficients of z^n, we find that

$$c_n = q^{2n-1} c_{n-1},$$

and so

$$c_n = q^{n^2} c_0 \tag{2.23}$$

for any positive integer n. Since $f(z) = f(z^{-1})$, the formula (2.23) holds for all negative integers n as well. We deduce that

$$\prod_{n=1}^{\infty} (1 + zq^{2n-1})(1 + z^{-1}q^{2n-1}) = c_0 \sum_{n=-\infty}^{\infty} q^{n^2} z^n,$$

where the constant term c_0 depends only on q and is independent of z. In order to complete this proof of (2.21), one must show that the constant term is given by

$$c_0 = \prod_{n=1}^{\infty} (1 - q^{2n})^{-1}.$$

There are no particular values of z for which the formula can be verified. Hence one must resort to a suitable artifice. In this connection, Ramanujan made a significant contribution. He generalized the product in (2.22) to include a denominator which consists of two infinite products.

Let

$$\Phi(z, \alpha, \beta) = \prod_{n=1}^{\infty} \frac{(1 + zq^{2n-1})(1 + z^{-1}q^{2n-1})}{(1 + \alpha zq^{2n-1})(1 + \beta z^{-1}q^{2n-1})}. \tag{2.24}$$

We regard Φ primarily as a function of z, with α, β and q as parameters. To begin with, we shall assume $\alpha\beta \neq 0$. The zeros and poles of Φ can be determined from the factors in the infinite product. The poles are given by

$$z \in \{-(\alpha q)^{-1}, -(\alpha q^3)^{-1}, -(\alpha q^5)^{-1}, \ldots\} \cup \{-\beta q, -\beta q^3, -\beta q^5, \ldots\}.$$

Suppose that $|\alpha\beta q^2| < 1$. The purpose of this condition is to ensure that $\Phi(z, \alpha, \beta)$ is analytic in the annular region

$$|\beta q| < |z| < |\alpha q|^{-1} \tag{2.25}$$

that separates the two infinite families of poles. By Laurent's theorem, Φ has an expansion of the form

$$\Phi(z, \alpha, \beta) = \sum_{n=-\infty}^{\infty} c_n(\alpha, \beta) z^n \tag{2.26}$$

in the annulus (2.25). One easily notices, using (2.24), that

$$\frac{\Phi(q^2z,\alpha,\beta)}{\Phi(z,\alpha,\beta)} = \frac{(1+\alpha qz)(1+q^{-1}z^{-1})}{(1+\beta q^{-1}z^{-1})(1+qz)},$$

or

$$(1+\alpha qz)\Phi(z,\alpha,\beta) = (\beta+qz)\Phi(q^2z,\alpha,\beta). \tag{2.27}$$

If we put

$$\Psi(z) = (1+\alpha qz)\Phi(z,\alpha,\beta),$$

we see that $\Psi(z)$ is analytic in the larger annulus

$$|\beta q| < |z| < |\alpha q^3|^{-1}, \tag{2.28}$$

because the pole of $\Phi(z,\alpha,\beta)$ at $z = -(\alpha q)^{-1}$ has been eliminated. Now

$$\Psi(z) = (1+\alpha qz)\sum_{n=-\infty}^{\infty} c_n(\alpha,\beta)z^n = \sum_{n=-\infty}^{\infty} (c_n(\alpha,\beta)+\alpha q c_{n-1}(\alpha,\beta))z^n,$$

and by (2.27),

$$\begin{aligned}\Psi(z) &= (\beta+qz)\sum_{n=-\infty}^{\infty} c_n(\alpha,\beta)q^{2n}z^n \\ &= \sum_{n=-\infty}^{\infty} (\beta c_n(\alpha,\beta)q^{2n} + c_{n-1}(\alpha,\beta)q^{2n-1})z^n.\end{aligned}$$

Both of these expansions for Ψ are valid in the larger annulus given by (2.28). Comparing coefficients of z^n gives

$$c_n(\alpha,\beta)+\alpha q c_{n-1}(\alpha,\beta) = \beta c_n(\alpha,\beta)q^{2n} + c_{n-1}(\alpha,\beta)q^{2n-1},$$

hence, we obtain the recurrence relation

$$c_n(\alpha,\beta) = \frac{q^{2n-2}-\alpha}{1-\beta q^{2n}}\, q c_{n-1}(\alpha,\beta).$$

Iterating, we obtain

$$c_n(\alpha,\beta) = \frac{(1-\alpha)(q^2-\alpha)\cdots(q^{2n-2}-\alpha)}{(1-\beta q^2)(1-\beta q^4)\cdots(1-\beta q^{2n})}\, q^n c_0(\alpha,\beta), \tag{2.29}$$

for any positive integer n. From the definition (2.24), it is clear that

$$\Phi(z,\alpha,\beta) = \Phi(z^{-1},\beta,\alpha)$$

and comparing coefficients of z^{-n} on each side gives

$$c_{-n}(\alpha,\beta) = c_n(\beta,\alpha). \tag{2.30}$$

Combining (2.29) and (2.30) it follows that for any positive integer n,

$$\begin{aligned} c_{-n}(\alpha,\beta) &= c_n(\beta,\alpha) \\ &= \frac{(1-\beta)(q^2-\beta)\cdots(q^{2n-2}-\beta)}{(1-\alpha q^2)(1-\alpha q^4)\cdots(1-\alpha q^{2n})} q^n c_0(\beta,\alpha) \\ &= \frac{(1-\beta)(q^2-\beta)\cdots(q^{2n-2}-\beta)}{(1-\alpha q^2)(1-\alpha q^4)\cdots(1-\alpha q^{2n})} q^n c_0(\alpha,\beta), \end{aligned} \tag{2.31}$$

where the last step follows by taking $n = 0$ in (2.30).

We still have to find the value of $c_0(\alpha,\beta)$. This can be done by finding the residue of $\Phi(z,\alpha,\beta)$ at the simple pole $z = -(\alpha q)^{-1}$. Now

$$(1+\alpha qz)\Phi(z,\alpha,\beta) = (1+\alpha qz)\sum_{n=0}^{\infty} c_n(\alpha,\beta)z^n + (1+\alpha qz)\sum_{n=1}^{\infty} c_{-n}(\alpha,\beta)z^{-n}. \tag{2.32}$$

By the ratio test, the series in (2.32) that involves negative powers of z converges for $|z| > |\beta q|$, hence

$$\lim_{z\to-(\alpha q)^{-1}} (1+\alpha qz)\sum_{n=1}^{\infty} c_{-n}(\alpha,\beta)z^{-n} = 0.$$

The series in (2.32) that involves non-negative powers of z is

$$(1+\alpha qz)\sum_{n=0}^{\infty} c_n(\alpha,\beta)z^n = c_0(\alpha,\beta) + \sum_{n=1}^{\infty}(\alpha q c_{n-1}(\alpha,\beta) + c_n(\alpha,\beta))z^n, \tag{2.33}$$

and by the ratio test, this converges for $|z| < |\alpha q^3|^{-1}$. Hence, to take the limit as $z \to -(\alpha q)^{-1}$, we need only substitute the value $z = -(\alpha q)^{-1}$ in the right hand side of (2.33). With this in mind, we take the limit as $z \to -(\alpha q)^{-1}$ in (2.32) and obtain

$$\begin{aligned} &\prod_{n=1}^{\infty} \frac{(1-\alpha^{-1}q^{2n-2})(1-\alpha q^{2n})}{(1-q^{2n})(1-\alpha\beta q^{2n})} \\ &= \lim_{N\to\infty}\left\{ c_0(\alpha,\beta) + \sum_{n=1}^{N}(\alpha q c_{n-1}(\alpha,\beta) + c_n(\alpha,\beta))(-\alpha q)^{-n} \right\} \\ &= \lim_{N\to\infty} c_N(\alpha,\beta)(-\alpha q)^{-N} \\ &= \lim_{N\to\infty} \frac{(1-\alpha^{-1})(1-\alpha^{-1}q^2)\cdots(1-\alpha^{-1}q^{2N-2})}{(1-\beta q^2)(1-\beta q^4)\cdots(1-\beta q^{2N})} c_0(\alpha,\beta) \\ &= c_0(\alpha,\beta)\prod_{n=1}^{\infty}\frac{(1-\alpha^{-1}q^{2n-2})}{(1-\beta q^{2n})}. \end{aligned}$$

Therefore,

$$c_0(\alpha, \beta) = \prod_{n=1}^{\infty} \frac{(1-\alpha q^{2n})(1-\beta q^{2n})}{(1-q^{2n})(1-\alpha\beta q^{2n})}. \tag{2.34}$$

Combining (2.24), (2.26), (2.29), (2.31) and (2.34), we obtain Ramanujan's formula:

$$\begin{aligned}\prod_{n=1}^{\infty} &\frac{(1+zq^{2n-1})(1+z^{-1}q^{2n-1})(1-q^{2n})(1-\alpha\beta q^{2n})}{(1+\alpha z q^{2n-1})(1+\beta z^{-1}q^{2n-1})(1-\alpha q^{2n})(1-\beta q^{2n})} \\ &= 1 + \sum_{n=1}^{\infty} \frac{(1-\alpha)(q^2-\alpha)\cdots(q^{2n-2}-\alpha)}{(1-\beta q^2)(1-\beta q^4)\cdots(1-\beta q^{2n})} q^n z^n \\ &\quad + \sum_{n=1}^{\infty} \frac{(1-\beta)(q^2-\beta)\cdots(q^{2n-2}-\beta)}{(1-\alpha q^2)(1-\alpha q^4)\cdots(1-\alpha q^{2n})} q^n z^{-n}.\end{aligned} \tag{2.35}$$

By the ratio test, the series involving positive powers of z in (2.35) converges for $|\alpha q z| < 1$, while the series involving negative powers of z converges for $|\beta q z^{-1}| < 1$. The two series will converge simultaneously and represent an analytic function of z when both conditions are satisfied, and this is precisely the annular region given by (2.25).

The formula (2.35) is called Ramanujan's ${}_1\psi_1$ (pronounced "one psi one") summation formula. G. H. Hardy [62, pp. 222 - 223] described it as *"a remarkable formula with many parameters"*. In fact it contains one more parameter than the classical identities of Jacobi. It is a wonder that Jacobi's rich collection of formulas does not contain this formula of Ramanujan.

If we take the limit as $\alpha,\ \beta \to 0$ in Ramanujan's formula (2.35), we obtain Jacobi's triple product identity (2.21):

$$\prod_{n=1}^{\infty}(1+zq^{2n-1})(1+z^{-1}q^{2n-1})(1-q^{2n}) = \sum_{n=-\infty}^{\infty} q^{n^2} z^n, \tag{2.36}$$

and the annulus of convergence (2.25) becomes $0 < |z| < \infty$.

In his second notebook, Ramanujan [92, Chapter 16, Entry 19] gives Jacobi's triple product identity in the form

$$1+\sum_{n=1}^{\infty}(ab)^{n(n-1)/2}(a^n+b^n) = \prod_{n=1}^{\infty}(1+a^{n-1}b^n)(1+a^n b^{n-1})(1-a^n b^n) \tag{2.37}$$

and points out that it is a special case of the ${}_1\psi_1$ summation formula (2.35). The formulas (2.36) and (2.37) may be seen to be equivalent by setting $q = (ab)^{1/2}$ and $z = (a/b)^{1/2}$ in (2.36).

In (2.36), replace q with $q^{1/2}$ and then replace z with $-q^{-1/2}z$ to obtain the familiar form

$$\prod_{n=1}^{\infty}(1-zq^{n-1})(1-z^{-1}q^n)(1-q^n)=\sum_{n=-\infty}^{\infty}(-1)^n q^{n(n-1)/2}z^n, \tag{2.38}$$

or

$$\prod_{n=1}^{\infty}(1-zq^{n})(1-z^{-1}q^n)(1-q^n)=\sum_{n=0}^{\infty}(-1)^n q^{n(n+1)/2}\frac{(z^{n+1/2}-z^{-n-1/2})}{(z^{1/2}-z^{-1/2})}. \tag{2.39}$$

Let $z \to 1$ in (2.39) to obtain

$$\prod_{n=1}^{\infty}(1-q^n)^3=\sum_{n=0}^{\infty}(-1)^n(2n+1)q^{n(n+1)/2}.$$

Now multiply by $q^{1/8}$ and complete the square in the exponent on the right hand side to get

$$q^{1/8}\prod_{n=1}^{\infty}(1-q^n)^3=\sum_{n=0}^{\infty}(-1)^n(2n+1)q^{(2n+1)^2/8}, \tag{2.40}$$

one of the most interesting identities of Jacobi, cf., (1.2).

Next, we replace q with q^3 in (2.38) and then put $z=q$ to obtain

$$\prod_{n=1}^{\infty}(1-q^{3n-2})(1-q^{3n-1})(1-q^{3n})=\sum_{n=-\infty}^{\infty}(-1)^n q^{n(3n-1)/2}.$$

We simplify the product on the left hand side to deduce Euler's product:

$$\prod_{n=1}^{\infty}(1-q^n)=\sum_{n=-\infty}^{\infty}(-1)^n q^{n(3n-1)/2}. \tag{2.41}$$

The identity (1.1) follows from this by multiplying by $q^{1/24}$ and completing the square in the exponent on the right hand side:

$$q^{1/24}\prod_{n=1}^{\infty}(1-q^n)=\sum_{n=-\infty}^{\infty}(-1)^n q^{(6n+1)^2/24}.$$

Ramanujan was able to add two more identities, cf., (1.3) and (1.4):

$$q^{1/24}\prod_{n=1}^{\infty}\frac{(1-q^n)^5}{(1-q^{2n})^2}=\sum_{n=-\infty}^{\infty}(6n+1)q^{(6n+1)^2/24}$$

and

$$q^{1/3}\prod_{n=1}^{\infty}\frac{(1-q^{2n})^5}{(1-(-q)^n)^2}=\sum_{n=-\infty}^{\infty}(3n+1)q^{(3n+1)^2/3}.$$

Proofs of these identities will be given in Appendix B.

Next, let $z=-1$ in (2.38) to get

$$2\prod_{n=1}^{\infty}(1+q^n)^2(1-q^n)=\sum_{n=-\infty}^{\infty}q^{n(n-1)/2}=2\sum_{n=1}^{\infty}q^{n(n-1)/2}.$$

By using the infinite product identity

$$\prod_{n=1}^{\infty}(1+q^n)=\prod_{n=1}^{\infty}\frac{(1-q^{2n})}{(1-q^n)}=\prod_{n=1}^{\infty}\frac{1}{(1-q^{2n-1})} \tag{2.42}$$

due to Euler, we obtain

$$\prod_{n=1}^{\infty}\frac{(1-q^{2n})^2}{(1-q^n)}=\sum_{n=0}^{\infty}q^{n(n+1)/2}, \tag{2.43}$$

a result first proved by Gauss.

If we take $z=-1$ in Jacobi's triple product identity (2.36), the resulting identity may be expressed in the form

$$\prod_{n=1}^{\infty}\frac{(1-q^n)^2}{(1-q^{2n})}=\sum_{n=-\infty}^{\infty}(-1)^n q^{n^2}. \tag{2.44}$$

On replacing q with $-q$ in (2.44) and noting that

$$\prod_{n=1}^{\infty}(1-(-q)^n)=\prod_{n=1}^{\infty}\frac{(1-q^{2n})^3}{(1-q^n)(1-q^{4n})},$$

we deduce that

$$\prod_{n=1}^{\infty}\frac{(1-q^{2n})^5}{(1-q^n)^2(1-q^{4n})^2}=\sum_{n=-\infty}^{\infty}q^{n^2}. \tag{2.45}$$

Here is another proof of Ramanujan's ${}_1\psi_1$ summation formula (2.35). We follow the first proof as far as (2.31). It remains to evaluate $c_0(\alpha,\beta)$. Again, it is easy to check from the definition (2.24) that

$$\Phi(z,\alpha q^2,\beta)=(1+\alpha qz)\Phi(z,\alpha,\beta). \tag{2.46}$$

Both sides of (2.46) are analytic in the larger annulus

$$|\beta q|<|z|<|\alpha q^3|^{-1}$$

because the pole of $\Phi(z,\alpha,\beta)$ at $z=-(\alpha q)^{-1}$ has been eliminated. Equating the constant coefficients and using (2.31) we obtain

$$c_0(\alpha q^2,\beta)=c_0(\alpha,\beta)+\alpha q c_{-1}(\alpha,\beta)=\frac{1-\alpha\beta q^2}{1-\alpha q^2}c_0(\alpha,\beta).$$

Iterating gives

$$c_0(\alpha,\beta) = c_0(\alpha q^{2n},\beta)\prod_{j=1}^{n}\frac{1-\alpha q^{2j}}{1-\alpha\beta q^{2j}},$$

and taking the limit as $n\to\infty$ gives

$$c_0(\alpha,\beta) = c_0(0,\beta)\prod_{j=1}^{\infty}\frac{1-\alpha q^{2j}}{1-\alpha\beta q^{2j}}. \tag{2.47}$$

By (2.30) and (2.47), we deduce further that

$$c_0(0,\beta) = c_0(\beta,0) = c_0(0,0)\prod_{j=1}^{\infty}(1-\beta q^{2j}), \tag{2.48}$$

and substituting (2.48) into (2.47) gives

$$c_0(\alpha,\beta) = c_0(0,0)\prod_{j=1}^{\infty}\frac{(1-\alpha q^{2j})(1-\beta q^{2j})}{1-\alpha\beta q^{2j}}. \tag{2.49}$$

Now take $\alpha=\beta=1$ and observe that $\Phi(z,1,1)=1$ identically. Hence, $c_0(1,1)=1$. Using this in (2.49) gives

$$1 = c_0(0,0)\prod_{j=1}^{\infty}(1-q^{2j})$$

and so

$$c_0(0,0) = \prod_{j=1}^{\infty}\frac{1}{(1-q^{2j})},$$

therefore (2.49) becomes

$$c_0(\alpha,\beta) = \prod_{j=1}^{\infty}\frac{(1-\alpha q^{2j})(1-\beta q^{2j})}{(1-\alpha\beta q^{2j})(1-q^{2j})}. \tag{2.50}$$

This completes the second proof of Ramanujan's ${}_1\psi_1$ summation formula.

2.3 Ramanujan's transcendentals U_{2n} and V_{2n}

Ramanujan has proved that if m and n are non-negative integers and $m+n$ is odd, then

$$\sum_{j=1}^{\infty}\sum_{k=1}^{\infty} j^m k^n q^{jk}$$

is a polynomial in P, Q and R. His notebooks contain two other transcendentals which are also polynomials in P, Q and R over the field of rational numbers. These are:

$$U_n = \left(1^{n+1} - 3^{n+1}q + 5^{n+1}q^3 - 7^{n+1}q^6 + \cdots\right) / \left(1 - 3q + 5q^3 - 7q^6 + \cdots\right)$$

and

$$V_n = \left(1^n - 5^n q - 7^n q^2 + 11^n q^5 + 13^n q^7 - \cdots\right)/(1 - q - q^2 + q^5 + q^7 - \cdots),$$

where the denominators are the Jacobi and Euler products (2.40) and (2.41), respectively. Ramanujan proved that

$$\begin{cases} U_0 = V_0 = 1, \\ U_{n+2} = PU_n + 8q\dfrac{dU_n}{dq}, \\ V_{n+2} = PV_n + 24q\dfrac{dV_n}{dq}, \end{cases} \tag{2.51}$$

and furthermore, that U_{2n} and V_{2n} are polynomials in P, Q and R over the field of rational numbers. In fact, the coefficients of the powers of P, Q and R in the expression for V_{2n} are integers, and those of U_{2r} are rational numbers whose denominators are powers of 3 only. Here we repeat the proof of (2.51) given by Ramanujan.

We multiply the numerator and denominator by $q^{1/8}$ and then apply Jacobi's formula (2.40) to write U_n in the form

$$U_n = \frac{\sum_{k=0}^{\infty}(-1)^k(2k+1)^{n+1}q^{(2k+1)^2/8}}{\sum_{k=0}^{\infty}(-1)^k(2k+1)q^{(2k+1)^2/8}} = \frac{\sum_{k=0}^{\infty}(-1)^k(2k+1)^{n+1}q^{(2k+1)^2/8}}{q^{1/8}\prod_{k=1}^{\infty}(1-q^k)^3}. \tag{2.52}$$

Now differentiate logarithmically with respect to q to obtain

$$\begin{aligned} 8q\frac{U_n'}{U_n} &= \frac{\sum_{k=0}^{\infty}(-1)^k(2k+1)^{n+3}q^{(2k+1)^2/8}}{\sum_{k=0}^{\infty}(-1)^k(2k+1)^{n+1}q^{(2k+1)^2/8}} - \left(1 - 24\sum_{k=1}^{\infty}\frac{kq^k}{1-q^k}\right) \\ &= \frac{U_{n+2}}{U_n} - P. \end{aligned}$$

Hence

$$U_{n+2} = PU_n + 8q\frac{dU_n}{dq}.$$

The proof for V_n is similar.

Ramanujan used the recurrence relations (2.51), and the differential equations for P, Q and R in (2.13), to successively calculate U_{2n} and V_{2n} for $1 \leq n \leq 6$. The results are given in Table 2.1.

Table 2.1 Ramanujan's transcendentals U_{2n} and V_{2n}

U_2	$=$	P
U_4	$=$	$(5P^2 - 2Q)/3$
U_6	$=$	$(35P^3 - 42PQ + 16R)/9$
U_8	$=$	$(35P^4 - 84P^2Q + 64PR - 12Q^2)/3$
U_{10}	$=$	$(385P^5 - 1540P^3Q + 1760P^2R - 660PQ^2 + 64RQ)/9$
U_{12}	$=$	$(5005P^6 - 30030P^4Q + 45760P^3R$ $-25740P^2Q^2 + 4992PQR + 552Q^3 - 512R^2)/27$
V_2	$=$	P
V_4	$=$	$3P^2 - 2Q$
V_6	$=$	$15P^3 - 30PQ + 16R$
V_8	$=$	$105P^4 - 420P^2Q + 448PR - 132Q^2$
V_{10}	$=$	$945P^5 - 6300P^3Q + 10080P^2R - 5940PQ^2 + 1216QR$
V_{12}	$=$	$10395P^6 - 103950P^4Q + 221760P^3R$ $-196020P^2Q^2 + 80256PQR - 2712Q^3 - 9728R^2.$

By induction, it can be shown that

$$U_{2n} = \sum_{\substack{j+2k+3\ell=n \\ j,k,\ell\geq 0}} \mu_{j,k,\ell} P^j Q^k R^\ell$$

and

$$V_{2n} = \sum_{\substack{j+2k+3\ell=n \\ j,k,\ell\geq 0}} \nu_{j,k,\ell} P^j Q^k R^\ell,$$

where the numbers $\mu_{j,k,\ell}$ are rational numbers whose denominators are powers of 3 only, and the coefficients $\nu_{j,k,\ell}$ are integers.

We remark that the transcendentals U_{2n} and V_{2n} are coefficients in the expansions of theta functions, and they are related to modular forms. In fact,

$$\sum_{n=0}^{\infty} U_{2n} \frac{(-1)^n \theta^{2n+1}}{(2n+1)!} = \sin\theta \prod_{n=1}^{\infty} \frac{(1-e^{2i\theta}q^n)(1-e^{-2i\theta}q^n)}{(1-q^n)^2} \tag{2.53}$$

and

$$\sum_{n=0}^{\infty} V_{2n} \frac{(-1)^n \theta^{2n}}{(2n)!} = \cos\theta \prod_{n=1}^{\infty} \frac{(1-e^{4i\theta}q^n)(1-e^{-4i\theta}q^n)}{(1-e^{2i\theta}q^n)(1-e^{-2i\theta}q^n)}. \tag{2.54}$$

The expansion (2.53) can be proved using Jacobi's triple product identity (2.38) or (2.39), while (2.54) is a consequence of the quintuple product identity that we will prove in Appendix 2.

2.4 The imaginary transformation and Dedekind's eta-function

In this section, suppose $\mathrm{Im}(\tau) > 0$ and let q and p be defined by

$$q = \exp(2\pi i\tau) \qquad \text{and} \qquad p = \exp(-2\pi i/\tau).$$

The functions $\phi_1(\theta)$ and $\wp(\theta)$ defined in Chapter 1 also depend on q, and hence on τ, and they will now be denoted by $\phi_1(\theta,\tau)$ and $\wp(\theta,\tau)$, respectively. From (1.28) and (1.36), we have

$$\wp(\theta,\tau) = -2\phi_1'(\theta,\tau) - \frac{P(q)}{12},$$

where the prime denotes differentiation with respect to θ. By the periodicity and quasi-periodicity properties of ϕ_1 given in (1.19) and (1.20), it follows that

$$\wp(\theta+2\pi,\tau) = \wp(\theta,\tau) = \wp(\theta+2\pi\tau,\tau),$$

that is, $\wp(\theta,\tau)$ is doubly periodic with periods 2π and $2\pi\tau$. Moreover, by considering the poles of ϕ_1 (see the line after (1.20)), it follows that $\wp(\theta,\tau)$ has poles of order 2 at each point $\theta = 2\pi m + 2\pi n\tau$ for all integers m and n, and no other singularities. The function $\wp(\theta/\tau, -1/\tau)$ has the same periods and poles as $\wp(\theta,\tau)$, and it is natural to expect that these functions must be intimately connected. Using the series expansion (1.38), we see that the difference

$$\wp(\theta,\tau)-\frac{1}{\tau^2}\wp\left(\frac{\theta}{\tau},-\frac{1}{\tau}\right) \tag{2.55}$$

$$=\frac{1}{240}\left(Q(q)-\frac{1}{\tau^4}Q(p)\right)\theta^2+\frac{1}{12\times 504}\left(R(q)-\frac{1}{\tau^6}R(p)\right)\theta^4+\cdots$$

is analytic at $\theta=0$, because the pole of $\wp(\theta,\tau)$ at $\theta=0$ has been eliminated. By double periodicity, the left hand side of (2.55) is analytic for all complex values of θ, and by Liouville's theorem it must be constant. By letting $\theta=0$ the constant is seen to be 0. Therefore, from (2.55) we have the transformation formula

$$\wp(\theta,\tau)=\frac{1}{\tau^2}\wp\left(\frac{\theta}{\tau},-\frac{1}{\tau}\right)$$

as well as the transformation formulas for the coefficients:

$$Q(q)=\frac{1}{\tau^4}Q(p),\qquad R(q)=\frac{1}{\tau^6}R(q),\qquad \text{etc.}$$

These transformation formulas follow easily from the double series definition of $\wp(\theta,\tau)$. However, the transformation of $P(q)$ is not so easy to derive since we meet non-absolutely convergent double series. This has been accomplished by a nice artifice by Serre [97, pp. 95–96]. The usual proofs make extensive use of contour integration. Ramanujan does not use the classical definition of $\wp(\theta)$, but has developed his own unique method of deriving the reciprocal formulas for P, Q, R, etc.

We may remark at this stage that the transformation formula for $P(q)$ may be proved using the elementary results of function theory, namely Liouville's theorem. The poles of $\phi_1(\theta,\tau)$ and $\tau^{-1}\phi_1(\theta/\tau,-1/\tau)$ occur at the same points $\theta=2\pi m+2\pi n\tau$, where m and n are any integers; they are all simple and have the same residue $1/2$. Therefore, the function g defined by

$$g(\theta)=\phi_1(\theta,\tau)-\frac{1}{\tau}\phi_1\left(\frac{\theta}{\tau},-\frac{1}{\tau}\right)$$

is entire. By the periodic and quasi-periodic properties (1.19) and (1.20) we see that

$$g(\theta+2\pi)=g(\theta)+\frac{1}{2i\tau}\qquad\text{and}\qquad g(\theta+2\pi\tau)=g(\theta)+\frac{1}{2i}.$$

It is easy to modify g to a doubly periodic function. In fact,

$$\phi_1(\theta,\tau)-\frac{1}{\tau}\phi_1\left(\frac{\theta}{\tau},-\frac{1}{\tau}\right)-\frac{\theta}{4\pi i\tau}$$

is entire and doubly periodic. By Liouville's theorem it is a constant function, and since ϕ_1 is an odd function, the constant is zero. Thus

$$\phi_1(\theta,\tau) = \frac{1}{\tau}\phi_1\left(\frac{\theta}{\tau}, -\frac{1}{\tau}\right) + \frac{\theta}{4\pi i\tau}. \tag{2.56}$$

Expanding in powers of θ using (1.33) and comparing coefficients, we again obtain the reciprocity formulas

$$Q(p) = \tau^4 Q(q) \qquad \text{and} \qquad R(p) = \tau^6 R(q) \tag{2.57}$$

as well as the more significant result

$$P(p) = \tau^2 P(q) + \frac{6\tau}{\pi i}, \tag{2.58}$$

where $q = e^{2\pi i\tau}$ and $p = e^{-2\pi i/\tau}$. Taking $\tau = i$ in (2.58) we get

$$P(e^{-2\pi}) = \frac{3}{\pi},$$

which, when written explicitly, gives a formula posed as a problem by Ramanujan:

$$\frac{1}{e^{2\pi}-1} + \frac{2}{e^{4\pi}-1} + \frac{3}{e^{6\pi}-1} + \cdots = \frac{1}{24} - \frac{1}{8\pi}. \tag{2.59}$$

We take this occasion to prove the fundamental result (2.56) by a simple method which does not make any use of complex function theory. It makes use of a natural duplication formula. By (1.11) and (1.15) we have

$$\phi_1(a,\tau) = \frac{1}{4i}\left(\frac{1+\alpha}{1-\alpha} + 2\sum_{n=1}^{\infty}\frac{\alpha q^n}{1-\alpha q^n} - \frac{\alpha^{-1}q^n}{1-\alpha^{-1}q^n}\right) \tag{2.60}$$

where $\alpha = e^{ia}$, $q = e^{2\pi i\tau}$. Replace a by $a+\pi$ (i.e., α by $-\alpha$) and add. The result is

$$\phi_1(a,\tau) + \phi_1(a+\pi,\tau) = \frac{1}{2i}\left(\frac{1+\alpha^2}{1-\alpha^2} + 2\sum_{n=1}^{\infty}\frac{\alpha^2 q^{2n}}{1-\alpha^2 q^{2n}} - \frac{\alpha^{-2}q^{2n}}{1-\alpha^{-2}q^{2n}}\right).$$

In this, replace a by $a+\pi\tau$ (i.e., α^2 by $q\alpha^2$) and add. The result simplifies to

$$\phi_1(a,\tau)+\phi_1(a+\pi,\tau)+\phi_1(a+\pi\tau,\tau)+\phi_1(a+\pi+\pi\tau,\tau) = -\frac{i}{2}+2\phi_1(2a,\tau). \tag{2.61}$$

By the quasi-periodic property (1.20) we also have

$$\phi_1(a,\tau)+\phi_1(a+\pi,\tau)+\phi_1(a-\pi\tau,\tau)+\phi_1(a+\pi-\pi\tau,\tau) = \frac{i}{2}+2\phi_1(2a,\tau). \tag{2.62}$$

Fix a value of τ and consider the function $\psi(a)$ defined by

$$\psi(a) = \tau\phi_1(a, \tau) - \phi_1\left(\frac{a}{\tau}, -\frac{1}{\tau}\right) - \frac{a}{4\pi i}. \tag{2.63}$$

By (2.61), (2.62) and (2.63), we find that

$$\psi(a) + \psi(a + \pi) + \psi(a + \pi\tau) + \psi(a + \pi + \pi\tau) = 2\psi(2a).$$

We have already seen that ψ is entire and doubly periodic with periods 2π and $2\pi\tau$. We will now show that ψ is identically zero, without using Cauchy's or Liouville's theorems. Differentiating with respect to a gives

$$\psi'(a) + \psi'(a + \pi) + \psi'(a + \pi\tau) + \psi'(a + \pi + \pi\tau) = 4\psi'(2a). \tag{2.64}$$

Let a_0 be a point in the parallelogram with vertices 0, 2π, $2\pi\tau$, $2\pi + 2\pi\tau$ for which $|\psi'(a) - \psi'(0)|$ is a maximum, say

$$|\psi'(a) - \psi'(0)| = M.$$

Then by (2.64),

$$\begin{aligned}
4M &= |4\psi'(a_0) - 4\psi'(0)| \\
&= \Big|\left(\psi'\left(\frac{a_0}{2}\right) - \psi'(0)\right) + \left(\psi'\left(\frac{a_0}{2} + \pi\right) - \psi'(0)\right) \\
&\qquad + \left(\psi'\left(\frac{a_0}{2} + \pi\tau\right) - \psi'(0)\right) + \left(\psi'\left(\frac{a_0}{2} + \pi + \pi\tau\right) - \psi'(0)\right)\Big| \\
&\le \left|\psi'\left(\frac{a_0}{2}\right) - \psi'(0)\right| + \left|\psi'\left(\frac{a_0}{2} + \pi\right) - \psi'(0)\right| \\
&\qquad + \left|\psi'\left(\frac{a_0}{2} + \pi\tau\right) - \psi'(0)\right| + \left|\psi'\left(\frac{a_0}{2} + \pi + \pi\tau\right) - \psi'(0)\right| \\
&\le 4M.
\end{aligned}$$

Hence the maximum of $|\psi'(a) - \psi'(0)|$ also occurs at $a_0/2$, and by the same argument at $a_0/2^n$, $n = 1, 2, 3, \ldots$. By continuity of ψ', the maximum occurs at $a = 0$. But $|\psi'(0) - \psi'(0)| = 0$, and so $\psi'(a) \equiv 0$. Thus ψ is constant, and since ψ is an odd function, $\psi(a) \equiv 0$. This again proves (2.56), this time without using complex function theory.

We use (2.60) to rewrite the transformation formula (2.56) in the explicit form

$$\begin{aligned}
&\frac{1}{4}\cot\frac{a}{2} + \frac{1}{2i}\sum_{n=1}^{\infty}\left(\frac{e^{ia}q^n}{1 - e^{ia}q^n} - \frac{e^{-ia}q^n}{1 - e^{-ia}q^n}\right) \\
&= \frac{1}{4\tau}\cot\frac{a}{2\tau} + \frac{1}{2i\tau}\sum_{n=1}^{\infty}\left(\frac{e^{ia/\tau}p^n}{1 - e^{ia/\tau}p^n} - \frac{e^{-ia/\tau}p^n}{1 - e^{-ia/\tau}p^n}\right) + \frac{a}{4\pi i\tau}.
\end{aligned}$$

Integrating both sides with repect to a, we obtain

$$\frac{1}{2}\log\sin\frac{a}{2}+\frac{1}{2}\sum_{n=1}^{\infty}\left(\log(1-e^{ia}q^n)+\log(1-e^{-ia}q^n)\right)$$

$$=\frac{1}{2}\log\sin\frac{a}{2\tau}+\frac{1}{2}\sum_{n=1}^{\infty}\left(\log(1-e^{ia/\tau}p^n)+\log(1-e^{-ia/\tau}p^n)\right)+\frac{a^2}{8\pi i\tau}+C_\tau,$$

for some constant C_τ that is independent of a. Now multiply by 2 and exponentiate, to get

$$\exp\left(\frac{-a^2}{4\pi i\tau}\right)\frac{\sin\frac{a}{2}}{\sin\frac{a}{2\tau}}\prod_{n=1}^{\infty}\frac{(1-e^{ia}q^n)(1-e^{-ia}q^n)}{(1-e^{ia/\tau}p^n)(1-e^{-ia/\tau}p^n)}=A_\tau \tag{2.65}$$

where A_τ is independent of a and depends only on τ. To determine A_τ, successively let a take the values π, $\pi\tau$ and $\pi+\pi\tau$ in (2.65), to get

$$A_\tau=2ip^{1/8}\prod_{n=1}^{\infty}\frac{(1+q^n)^2}{(1-p^{n-1/2})^2}=\frac{i}{2q^{1/8}}\prod_{n=1}^{\infty}\frac{(1-q^{n-1/2})^2}{(1+p^n)^2} \tag{2.66}$$

$$=i\left(\frac{p}{q}\right)^{1/8}\prod_{n=1}^{\infty}\frac{(1+q^{n-1/2})^2}{(1+p^{n-1/2})^2}.$$

We recall Euler's identity (2.42)

$$\prod_{n=1}^{\infty}(1+q^n)(1-q^{2n-1})=1,$$

and separate the terms in the first factor according to whether n is even or odd, respectively, to get

$$\prod_{n=1}^{\infty}(1+q^{2n})(1+q^{2n-1})(1-q^{2n-1})=1. \tag{2.67}$$

Now multiply the three expressions in (2.66) and make use of (2.67) to get

$$A_\tau^3=i^3\left(\frac{p}{q}\right)^{1/4}.$$

Hence,

$$A_\tau=\omega\, i\left(\frac{p}{q}\right)^{1/12}=\omega\, i\exp\left(-\frac{\pi i}{6}\left(\tau+\frac{1}{\tau}\right)\right) \tag{2.68}$$

for some cube root of unity ω. We now put $a=0$ in (2.65) to get

$$A_\tau=\tau\prod_{n=1}^{\infty}\frac{(1-q^n)^2}{(1-p^n)^2}. \tag{2.69}$$

For $\mathrm{Im}(\tau) > 0$, A_τ is an analytic function of τ, and when $\tau = i$ we have $q = p = e^{-2\pi}$. Hence, from (2.69) we get $A_i = i$. Therefore we deduce that $\omega = 1$ in (2.68), and

$$A_\tau = i\left(\frac{p}{q}\right)^{1/12} = i\exp\left(-\frac{\pi i}{6}\left(\tau + \frac{1}{\tau}\right)\right). \tag{2.70}$$

Using the value (2.70) in each of the identities (2.66) and (2.69), we deduce

$$q^{1/24}\prod_{n=1}^{\infty}(1+q^n) = \frac{1}{\sqrt{2}\,p^{1/48}}\prod_{n=1}^{\infty}(1-p^{n-1/2}), \tag{2.71}$$

$$\frac{1}{q^{1/48}}\prod_{n=1}^{\infty}(1-q^{n-1/2}) = \sqrt{2}\,p^{1/24}\prod_{n=1}^{\infty}(1+p^n), \tag{2.72}$$

$$\frac{1}{q^{1/48}}\prod_{n=1}^{\infty}(1+q^{n-1/2}) = \frac{1}{\sqrt{2}\,p^{1/48}}\prod_{n=1}^{\infty}(1+p^{n-1/2}) \tag{2.73}$$

and

$$q^{1/24}\prod_{n=1}^{\infty}(1-q^n) = \sqrt{\frac{i}{\tau}}\,p^{1/24}\prod_{n=1}^{\infty}(1-p^n), \tag{2.74}$$

where $q = \exp(2\pi i\tau)$, $p = \exp(-2\pi i/\tau)$ and $\mathrm{Im}(\tau) > 0$. The last formula is the classical transformation for Dedekind's eta-function. The branch of the square root is determined by requiring that $\sqrt{i/\tau}$ be positive when τ is purely imaginary.

In Ramanujan's ${}_1\psi_1$ summation formula (2.35), take $\beta = 1/\alpha$ and divide both sides by $1-\alpha$, to get

$$\begin{aligned}
&\prod_{n=1}^{\infty}\frac{(1+zq^{2n-1})(1+z^{-1}q^{2n-1})(1-q^{2n})^2}{(1+\alpha zq^{2n-1})(1+\alpha^{-1}z^{-1}q^{2n-1})(1-\alpha q^{2n-2})(1-\alpha^{-1}q^{2n})}\\
&= \frac{1}{1-\alpha} + \sum_{n=1}^{\infty}\frac{(-1)^n}{q^{2n}-\alpha}(\alpha qz)^n + \sum_{n=1}^{\infty}\frac{(-1)^n}{1-\alpha q^{2n}}\left(\frac{q}{\alpha z}\right)^n\\
&= \sum_{n=-\infty}^{\infty}\frac{1}{1-\alpha q^{2n}}\left(-\frac{q}{\alpha z}\right)^n,
\end{aligned}$$

valid in the annulus $|q| < |\alpha z| < |q|^{-1}$. The series may be simplified by making the change of variable $t = -q/\alpha z$, then replacing q^2 with q. The result is

$$\prod_{n=1}^{\infty}\frac{(1-\alpha tq^{n-1})(1-\alpha^{-1}t^{-1}q^n)(1-q^n)^2}{(1-tq^{n-1})(1-tq^n)(1-\alpha q^{n-1})(1-\alpha^{-1}q^n)} = \sum_{n=-\infty}^{\infty}\frac{t^n}{1-\alpha q^n}, \tag{2.75}$$

provided $|q| < |t| < 1$. The product in (2.75) is symmetric in α and t, but the series is not. Symmetry can be restored to the series as follows:

$$\begin{aligned}\sum_{n=-\infty}^{\infty} \frac{t^n}{1-\alpha q^n} &= \frac{1}{1-\alpha} + \sum_{n=1}^{\infty} \frac{t^n(1-\alpha q^n + \alpha q^n)}{1-\alpha q^n} + \sum_{n=1}^{\infty} \frac{t^{-n}}{1-\alpha q^{-n}} \\ &= \frac{1}{1-\alpha} + \frac{t}{1-t} + \sum_{n=1}^{\infty} \frac{\alpha t^n q^n}{1-\alpha q^n} - \sum_{n=1}^{\infty} \frac{\alpha^{-1} t^{-n} q^n}{1-\alpha^{-1} q^n} \\ &= \frac{1}{1-\alpha} + \frac{1}{1-t} - 1 + \sum_{n=1}^{\infty}\sum_{m=1}^{\infty} q^{mn}(\alpha^m t^n - \alpha^{-m} t^{-n}).\end{aligned}$$

Thus,

$$\prod_{n=1}^{\infty} \frac{(1-\alpha t q^{n-1})(1-\alpha^{-1}t^{-1}q^n)(1-q^n)^2}{(1-tq^{n-1})(1-tq^n)(1-\alpha q^{n-1})(1-\alpha^{-1}q^n)} \tag{2.76}$$

$$= \frac{1}{1-\alpha} + \frac{1}{1-t} - 1 + \sum_{n=1}^{\infty}\sum_{m=1}^{\infty} q^{mn}(\alpha^m t^n - \alpha^{-m} t^{-n}),$$

provided $|q| < |\alpha|,\ |t| < |q|^{-1}$ and $\alpha,\ t \neq 1$. This function was studied by Jordan and Kronecker. We shall take up this topic in the next chapter where we consider in detail another interesting generalization of Ramanujan's identity (1.5).

2.5 Notes

The material in Sec. 2.1 is based on Secs. 5, 6, 7 and 10 of Ramanujan's paper [91]. The differential equations (2.13) have come to be known as Ramanujan's differential equations, although they were known earlier to G.-H. Halphen in 1886 [60, p. 450]. For a different, recent proof, and references to other proofs, see the paper by H. H. Chan [39].

Analogues of Ramanujan's differential equations for other levels have been given by V. Ramamani [87, pp. 112–123], [88], H. Hahn [59], R. S. Maier [84], T. Huber [66], [67], [68] and P. C. Toh [100].

The congruence (2.17) was given by Ramanujan in a manuscript that appears in his Lost Notebook [93, pp. 153, 168, 174]. A redaction of this manuscript has been published by B. C. Berndt and K. Ono [23]. Other interesting congruences and convolution sums for Ramanujan's tau-function have been given in the work by B. Ramakrishnan and B. Sahu [86].

Ramanujan's ${}_1\psi_1$ summation formula (2.35) appears in Ramanujan's second notebook [92, Chapter 16, Entry 17]. For some historical background, proofs and references, see the work by C. Adiga, B. C. Berndt,

S. Bhargava and G. N. Watson [1], the book by G. E. Andrews and B. C. Berndt [4, pp. 54–56] and the papers by R. A. Askey [6] and W. P. Johnson [74].

Ramanujan's transcendentals U_{2n} and V_{2n} are in the lost notebook [93, p. 369]. In his presidential address delivered to the Indian Mathematical Society [104], Venkatachaliengar described the proof of (2.51) outlined by Ramanujan in the lost notebook [93, p. 369] as "disarmingly simple". The identities in (2.51) have been discussed further by B. C. Berndt and A. J. Yee [26]; see also in [4, pp. 355–364]. Analogues of (2.51) for which the product in the denominator of (2.52) is replaced with $q^{r/24}\prod_{\ell=1}^{\infty}(1-q^{\ell})^r$ for $r \in \{2,4,6,8,10,14,26\}$ have been discovered by H. H. Chan, S. Cooper and P. C. Toh [40]. Further analogues appear in the papers by H. Hahn [59], T. Huber [66], [67] and Toh [100].

The identities (2.53) and (2.54) were given by Chan [39, (2.3) and (3.2)].

The identity (2.59) was posed as a problem by Ramanujan [94, p. 326, Question 387 and pp. 392–393] and recorded in his second notebook [92, p. 170]. A discussion and further references have been given by Berndt [9].

The identity (2.65), together with the value of A_τ given by (2.70), is the classical transformation formula for the theta function. Standard methods of proof include using the Poisson summation formula (e.g., see [3, p. 119, Ex. 30]) or analyzing the zeros and using Liouville's theorem (e.g., see [103] or [111, p. 475]). A variation of the proof given above has been given in [45]. A completely different proof has been given recently by Berndt, C. Gugg, S. Konsiriwong and J. Thiel [22].

Chapter 3

The Jordan-Kronecker Function

3.1 The Jordan-Kronecker function

At the end of the previous chapter, we have obtained the infinite product formula (2.75) for the Jordan-Kronecker function $f(\alpha, t)$, defined by

$$f(\alpha, t) = \sum_{n=-\infty}^{\infty} \frac{t^n}{1 - \alpha q^n}. \tag{3.1}$$

This is a generalization of the Ramanujan function ρ_1 defined in (1.8). We will prove that $f(\alpha, t)$ satisfies an identity of the same type as (1.9), namely:

$$f(\alpha, t)f(\beta, t) = t\frac{\partial}{\partial t} f(\alpha\beta, t) + (\rho_1(\alpha) + \rho_1(\beta))\, f(\alpha\beta, t).$$

Taking $t = e^v$, this can also be written in the form

$$f(\alpha, e^v)f(\beta, e^v) = \frac{\partial}{\partial v} f(\alpha\beta, e^v) + (\rho_1(\alpha) + \rho_1(\beta))\, f(\alpha\beta, e^v).$$

From this we shall derive the interesting identity

$$\begin{aligned} &\sum_{r=0}^{m} \binom{m}{r} \rho_{r+1}(\alpha)\rho_{m+1-r}(\beta) \\ &= \frac{\rho_{m+2}(\alpha) + \rho_{m+2}(\beta)}{m+1} + \rho_{m+2}(\alpha\beta) + (\rho_1(\alpha) + \rho_1(\beta))\, \rho_{m+1}(\alpha\beta), \end{aligned} \tag{3.2}$$

which holds for any non-negative integer m. Here $\rho_1(z)$ is the generalized Ramanujan function given by (1.8), or by its analytic continuation (1.11):

$$\rho_1(z) = \frac{1+z}{2(1-z)} + \sum_{n=1}^{\infty} \left(\frac{zq^n}{1 - zq^n} - \frac{z^{-1}q^n}{1 - z^{-1}q^n} \right);$$

and for $r \geq 2$, $\rho_r(z)$ is defined by

$$\rho_r(z) = -\frac{B_r}{r} + \sum_{n=1}^{\infty} n^{r-1} \left(\frac{zq^n}{1 - zq^n} + (-1)^r \frac{z^{-1}q^n}{1 - z^{-1}q^n} \right), \tag{3.3}$$

and B_r is the Bernoulli number defined by (1.30). The case $r = 2$ in (3.3) coincides with the function $\rho_2(z)$ studied earlier in (1.12). Therefore when $m = 0$, the identity (3.2) is just the generalized Ramanujan identity (1.9).

We deduced the product formula for $f(\alpha, t)$ at the end of the previous chapter as a special case of Ramanujan's ${}_1\psi_1$ summation formula. We shall now study further properties of $f(\alpha, t)$ that will be required for our subsequent work.

We begin with the formula (2.76):

$$f(\alpha,t) = \frac{1}{1-\alpha} + \frac{1}{1-t} - 1 + \sum_{n=1}^{\infty}\sum_{m=1}^{\infty} q^{mn}(\alpha^m t^n - \alpha^{-m}t^{-n}), \tag{3.4}$$

the series on the right hand side being valid for $|q| < |\alpha|$, $|t| < |q|^{-1}$. With this restriction, we have the interesting formulas

$$f(\alpha,t) = f(t,\alpha) \qquad \text{and} \qquad f(\alpha^{-1},t^{-1}) = -f(\alpha,t). \tag{3.5}$$

We now derive a series representation due to C. Jordan. We break the terms in the series in (3.4) into three cases: $m = n$, $m = n + r$ and $n = m + r$, where r is a positive integer, to get

$$\sum_{m=1}^{\infty}\sum_{n=1}^{\infty} q^{mn}\alpha^m t^n$$
$$= \sum_{n=1}^{\infty} q^{n^2}\alpha^n t^n + \sum_{n=1}^{\infty}\sum_{r=1}^{\infty} q^{(n+r)n}\alpha^{n+r}t^n + \sum_{m=1}^{\infty}\sum_{r=1}^{\infty} q^{m(m+r)}\alpha^m t^{m+r}.$$

Summing over r, we get

$$\sum_{m=1}^{\infty}\sum_{n=1}^{\infty} q^{mn}\alpha^m t^n = \sum_{n=1}^{\infty} q^{n^2}\alpha^n t^n\left(1 + \frac{\alpha q^n}{1-\alpha q^n} + \frac{tq^n}{1-tq^n}\right). \tag{3.6}$$

Using (3.6) in (3.4), we obtain Jordan's interesting formula

$$\begin{aligned} f(\alpha,t) = {} & \frac{1-\alpha t}{(1-\alpha)(1-t)} \\ & + \sum_{n=1}^{\infty} q^{n^2}\alpha^n t^n\left(1 + \frac{\alpha q^n}{1-\alpha q^n} + \frac{tq^n}{1-tq^n}\right) \\ & - \sum_{n=1}^{\infty} q^{n^2}\alpha^{-n} t^{-n}\left(1 + \frac{\alpha^{-1} q^n}{1-\alpha^{-1} q^n} + \frac{t^{-1}q^n}{1-t^{-1}q^n}\right), \end{aligned} \tag{3.7}$$

which holds for all values of α and t, except for $\alpha = t^n$, $t = q^m$, for integers m and n, where there are poles. The formula (3.7) shows that the identities in (3.5) are valid for all values of α and t.

We now derive the functional equation

$$f(\alpha, t) = tf(\alpha q, t) \tag{3.8}$$

from Jordan's formula (3.7). First of all, we observe that (3.7) may be written as

$$f(\alpha, t) = \sum_{n=-\infty}^{\infty} q^{n^2} \alpha^n t^n \left(\frac{1}{1-\alpha q^n} + \frac{1}{1-tq^n} - 1 \right).$$

Hence,

$$\begin{aligned} tf(\alpha q, t) &= \sum_{n=-\infty}^{\infty} q^{n^2+n} \alpha^n t^{n+1} \left(\frac{1}{1-\alpha q^{n+1}} + \frac{1}{1-tq^n} - 1 \right) \qquad (3.9) \\ &= \sum_{n=-\infty}^{\infty} q^{(n+1)^2} \alpha^{n+1} t^{n+1} \left(\frac{\alpha^{-1} q^{-n-1}}{1-\alpha q^{n+1}} \right) \\ &\quad + \sum_{n=-\infty}^{\infty} q^{n^2} \alpha^n t^n \left(\frac{tq^n}{1-tq^n} \right) - \sum_{n=-\infty}^{\infty} q^{n^2+n} \alpha^n t^{n+1}. \end{aligned}$$

The first series on the right hand side of (3.9) is

$$\begin{aligned} &\sum_{n=-\infty}^{\infty} q^{(n+1)^2} \alpha^{n+1} t^{n+1} \left(\frac{\alpha^{-1} q^{-n-1} - 1 + 1}{1-\alpha q^{n+1}} \right) \\ &= \sum_{n=-\infty}^{\infty} q^{(n+1)^2} \alpha^{n+1} t^{n+1} \left(\frac{1}{1-\alpha q^{n+1}} \right) \\ &\qquad + \sum_{n=-\infty}^{\infty} q^{(n+1)^2} \alpha^{n+1} t^{n+1} q^{-n-1} \alpha^{-1} \\ &= \sum_{n=-\infty}^{\infty} q^{n^2} \alpha^n t^n \left(\frac{1}{1-\alpha q^n} \right) + \sum_{n=-\infty}^{\infty} q^{n^2+n} \alpha^n t^{n+1}. \end{aligned}$$

Substitute this back into (3.9) to get

$$\begin{aligned} tf(\alpha q, t) &= \sum_{n=-\infty}^{\infty} q^{n^2} \alpha^n t^n \left(\frac{1}{1-\alpha q^n} \right) + \sum_{n=-\infty}^{\infty} q^{n^2} \alpha^n t^n \left(\frac{tq^n}{1-tq^n} \right) \\ &= \sum_{n=-\infty}^{\infty} q^{n^2} \alpha^n t^n \left(\frac{1}{1-\alpha q^n} + \frac{1}{1-tq^n} - 1 \right) \\ &= f(\alpha, t). \end{aligned}$$

In Chapter 2 we deduced the product formula (2.75) for $f(\alpha, t)$ from Ramanujan's ${}_1\psi_1$ summation formula (2.35). The symmetry properties

(3.5) and the functional equation (3.8) follow immediately from the product formula (2.75). We have shown that (3.5) and (3.8) may also be derived from Jordan's formula (3.7) on account of its intrinsic interest.

We now give another proof of the product formula (2.75) in a simple way. Let us temporarily define

$$\theta(z)=\prod_{n=1}^{\infty}(1-zq^{n-1})(1-z^{-1}q^n),$$

valid for $0<|z|<\infty$. One easily verifies

$$\theta(qz)=-z^{-1}\theta(z). \tag{3.10}$$

Consider the function $N(\alpha,t)$ defined by

$$N(\alpha,t)=f(\alpha,t)\theta(\alpha)\theta(t), \tag{3.11}$$

where f is the series given by Jordan's formula (3.7). At each of the simple poles of $f(\alpha,t)$, the function $\theta(\alpha)\theta(t)$ has a simple zero. Hence, $N(\alpha,t)$ is analytic in $0<|\alpha|,\ |t|<\infty$ and possesses a Laurent expansion in this region. From the functional equations (3.8) and (3.10), we obtain

$$N(\alpha q,t)=-\alpha^{-1}t^{-1}N(\alpha,t)=N(\alpha,tq).$$

Now $\theta(\alpha t)$ also has a Laurent expansion in the same annulus with the same recurrence formula. Hence we obtain by elementary reasoning

$$N(\alpha,t)=\theta(\alpha t)\lambda(q). \tag{3.12}$$

where $\lambda(q)$ is independent of α and t. In order to determine $\lambda(q)$, we take the limit as $t\to 1$ to get

$$\begin{aligned}\lambda(q)&=\lim_{t\to 1}\frac{N(\alpha,t)}{\theta(\alpha t)} \qquad (3.13)\\ &=\lim_{t\to 1}\frac{f(\alpha,t)\theta(\alpha)\theta(t)}{\theta(\alpha t)}\\ &=\lim_{t\to 1}(1-t)f(\alpha,t)\prod_{n=1}^{\infty}(1-tq^n)(1-t^{-1}q^n)\\ &=\prod_{n=1}^{\infty}(1-q^n)^2.\end{aligned}$$

Hence, combining (3.11), (3.12) and (3.13), we obtain finally

$$\begin{aligned}f(\alpha,t)&=\frac{\theta(\alpha,t)}{\theta(\alpha)\theta(t)}\prod_{n=1}^{\infty}(1-q^n)^2 \qquad (3.14)\\ &=\prod_{n=1}^{\infty}\frac{(1-\alpha tq^{n-1})(1-\alpha^{-1}t^{-1}q^n)(1-q^n)^2}{(1-tq^{n-1})(1-t^{-1}q^n)(1-\alpha q^{n-1})(1-\alpha^{-1}q^n)},\end{aligned}$$

which we obtained at the end of Chapter 2 using Ramanujan's ${}_1\psi_1$ summation formula.

3.2 The fundamental multiplicative identity

We now proceed to prove the identity

$$f(\alpha,t)f(\beta,t) = \frac{\partial}{\partial t} f(q^{-1}\alpha\beta,t) + t^{-1}\left(\rho_1(\alpha)+\rho_1(\beta)-1\right) f(q^{-1}\alpha\beta,t). \tag{3.15}$$

By (1.8) we have

$$\rho_1(z) = \frac{1}{2} + \sum_n{}' \frac{z^n}{1-q^n}$$

where the prime denotes that the summation is over all non-zero integers n. Furthermore, by the definition (3.1) and the symmetry property (3.5), we have

$$f(\alpha,t) = \sum_{n=-\infty}^{\infty} \frac{\alpha^n}{1-tq^n}.$$

Therefore the terms in (3.15) have expansions given by

$$f(\alpha,t)f(\beta,t) = \sum_{m=-\infty}^{\infty}\sum_{n=-\infty}^{\infty} \frac{\alpha^n\beta^m}{(1-tq^n)(1-tq^m)},$$

$$\frac{\partial}{\partial t} f(q^{-1}\alpha\beta,t) = \sum_{n=-\infty}^{\infty} \frac{\alpha^n\beta^n}{(1-tq^n)^2},$$

$$t^{-1}\left(\rho_1(\alpha)-1/2\right) f(q^{-1}\alpha\beta,t) = \sum_r{}' \frac{\alpha^r t^{-1}}{1-q^r} \sum_{m=-\infty}^{\infty} \frac{\alpha^m\beta^m q^{-m}}{1-tq^m},$$

and

$$t^{-1}\left(\rho_1(\beta)-1/2\right) f(\alpha\beta q^{-1},t) = \sum_r{}' \frac{\beta^r t^{-1}}{1-q^r} \sum_{m=-\infty}^{\infty} \frac{\alpha^m\beta^m q^{-m}}{1-tq^m},$$

where the series converge for

(i) $|q| < |\alpha|,\ |\beta| < 1$,
(ii) $|q|^2 < |\alpha\beta| < 1$,
(iii) $|q| < |\alpha| < 1$ and $|q|^2 < |\alpha\beta| < |q|$,
(iv) $|q| < |\beta| < 1$ and $|q|^2 < |\alpha\beta| < |q|$,

respectively. All of the conditions (i)–(iv) will be satisfied if

$$|q| < |\alpha|,\ |\beta| < |q|^{1/2}, \tag{3.16}$$

which we temporarily assume. We also assume that $t \neq q^r$ for any integer r. Now we observe that the coefficient of $\alpha^n \beta^n$ on either side of (3.15) is $1/(1-tq^n)^2$. Also, for $r \neq 0$ the coefficient of $\alpha^{n+r}\beta^n$ on the right hand side of (3.15) is

$$\frac{t^{-1}q^{-n}}{(1-q^r)(1-tq^n)} + \frac{t^{-1}q^{-n-r}}{(1-q^{-r})(1-tq^{n+r})} = \frac{1}{(1-tq^n)(1-tq^{n+r})}$$

and this is the same as the coefficient of the corresponding term on the left hand side of (3.15). Therefore the identity (3.15) is proved, subject to the condition (3.16) which can be removed by analytic continuation or appropriate functional equations.

By the functional equation (3.8) we have

$$\begin{aligned}\frac{\partial}{\partial t} f(q^{-1}\alpha\beta, t) &= \frac{\partial}{\partial t}\left(t f(\alpha\beta, t)\right)\\ &= t\frac{\partial}{\partial t} f(\alpha\beta, t) + f(\alpha\beta, t)\\ &= t\frac{\partial}{\partial t} f(\alpha\beta, t) + t^{-1} f(q^{-1}\alpha\beta, t),\end{aligned}$$

therefore the identity (3.15) may be written in the simpler form

$$f(\alpha, t) f(\beta, t) = t\frac{\partial}{\partial t} f(\alpha\beta, t) + \left(\rho_1(\alpha) + \rho_1(\beta)\right) f(\alpha\beta, t). \tag{3.17}$$

We shall refer to (3.17) as the Fundamental Multiplicative Identity.

By series manipulations, we have

$$\begin{aligned}f(\alpha, t) &= \sum_{n=-\infty}^{\infty} \frac{t^n}{1-\alpha q^n}\\ &= \frac{1}{1-\alpha} + \sum_{n=1}^{\infty} \frac{t^n(1-\alpha q^n + \alpha q^n)}{1-\alpha q^n} + \sum_{n=1}^{\infty} \frac{t^{-n}}{1-\alpha q^{-n}}\\ &= \frac{1}{1-\alpha} + \frac{1}{1-t} - 1 + \sum_{n=1}^{\infty} \frac{\alpha q^n t^n}{1-\alpha q^n} - \sum_{n=1}^{\infty} \frac{\alpha^{-1} q^n t^{-n}}{1-\alpha^{-1} q^n}.\end{aligned}$$

We put $t = e^v$ and expand in powers of v to obtain

$$f(\alpha, e^v) = \frac{1}{1-\alpha} + \frac{1}{1-e^v} - 1 + \sum_{n=1}^{\infty}\left(\frac{\alpha q^n e^{nv}}{1-\alpha q^n} - \frac{\alpha^{-1}q^n e^{-nv}}{1-\alpha^{-1}q^n}\right) \tag{3.18}$$

$$= \frac{1}{1-\alpha} - 1 - \sum_{m=0}^{\infty}\frac{B_m v^{m-1}}{m!}$$

$$+ \sum_{n=1}^{\infty}\left(\frac{\alpha q^n}{1-\alpha q^n}\sum_{m=0}^{\infty}\frac{(nv)^m}{m!} - \frac{\alpha^{-1}q^n}{1-\alpha^{-1}q^n}\sum_{m=0}^{\infty}\frac{(-nv)^m}{m!}\right)$$

$$= -\frac{1}{v} + \left(\frac{1+\alpha}{2(1-\alpha)} + \sum_{n=1}^{\infty}\frac{\alpha q^n}{1-\alpha q^n} - \frac{\alpha^{-1}q^n}{1-\alpha^{-1}q^n}\right)$$

$$+ \sum_{m=1}^{\infty}\left(-\frac{B_{m+1}}{m+1} + \sum_{n=1}^{\infty}\frac{n^m \alpha q^n}{1-\alpha q^n} - \frac{(-1)^m n^m \alpha^{-1}q^n}{1-\alpha^{-1}q^n}\right)\frac{v^m}{m!}$$

$$= -\frac{1}{v} + \sum_{m=0}^{\infty}\rho_{m+1}(\alpha)\frac{v^m}{m!},$$

where

$$\rho_1(\alpha) = \frac{1+\alpha}{2(1-\alpha)} + \sum_{n=1}^{\infty}\left(\frac{\alpha q^n}{1-\alpha q^n} - \frac{\alpha^{-1}q^n}{1-\alpha^{-1}q^n}\right) \tag{3.19}$$

and for $m = 2, 3, 4, \ldots,$

$$\rho_m(\alpha) = -\frac{B_m}{m} + \sum_{n=1}^{\infty}\left(\frac{n^{m-1}\alpha q^n}{1-\alpha q^n} + (-1)^m\frac{n^{m-1}\alpha^{-1}q^n}{1-\alpha^{-1}q^n}\right). \tag{3.20}$$

The functions ρ_1 and ρ_2 are exactly the same as the ones already encountered in (1.11) and (1.12). For $m = 3, 4, 5, \ldots,$ we take (3.20) as the definition of ρ_m.

Under the change of variable $t = e^v$, the Fundamental Multiplicative Identity (3.17) becomes

$$f(\alpha, e^v)f(\beta, e^v) = \frac{\partial}{\partial v}f(\alpha\beta, e^v) + f(\alpha\beta, e^v)(\rho_1(\alpha) + \rho_1(\beta)).$$

We expand both sides in powers of v, using (3.18), to get

$$\left(-\frac{1}{v} + \sum_{m=0}^{\infty}\rho_{m+1}(\alpha)\frac{v^m}{m!}\right)\left(-\frac{1}{v} + \sum_{m=0}^{\infty}\rho_{m+1}(\beta)\frac{v^m}{m!}\right)$$

$$= \left(\frac{1}{v^2} + \sum_{m=0}^{\infty}\rho_{m+2}(\alpha\beta)\frac{v^m}{m!}\right)$$

$$+ \left(-\frac{1}{v} + \sum_{m=0}^{\infty}\rho_{m+1}(\alpha\beta)\frac{v^m}{m!}\right)(\rho_1(\alpha) + \rho_1(\beta)).$$

The coefficients of v^{-2} and v^{-1} on both sides are equal. Equating coefficients of v^m for $m = 0, 1, 2, \ldots$, gives

$$\sum_{k=0}^{m} \binom{m}{k} \rho_{k+1}(\alpha)\rho_{m+1-k}(\beta) \tag{3.21}$$
$$= \frac{\rho_{m+2}(\alpha) + \rho_{m+2}(\beta)}{m+1} + \rho_{m+2}(\alpha\beta) + \rho_{m+1}(\alpha\beta)(\rho_1(\alpha) + \rho_1(\beta)).$$

When $m = 0$ this reduces to

$$\rho_1(\alpha)\rho_1(\beta) = \rho_2(\alpha) + \rho_2(\beta) + \rho_2(\alpha\beta) + \rho_1(\alpha\beta)\left(\rho_1(\alpha) + \rho_1(\beta)\right),$$

which is a restatement of the generalized Ramanujan identity (1.9). We rewrite (3.21) in the form

$$\sum_{k=0}^{m-1} \binom{m}{k} \rho_{k+1}(\alpha)\rho_{m+1-k}(\beta)$$
$$= \frac{\rho_{m+2}(\alpha) + \rho_{m+2}(\beta)}{m+1}$$
$$+\rho_{m+2}(\alpha\beta) + \rho_{m+1}(\alpha\beta)\rho_1(\alpha) + \left(\frac{\rho_{m+1}(\alpha\beta) - \rho_{m+1}(\alpha)}{1-\beta}\right)(1-\beta)\rho_1(\beta)$$

and take the limit as $\beta \to 1$, to get

$$\sum_{k=0}^{m-1} \binom{m}{k} \rho_{k+1}(\alpha)\rho_{m+1-k}(1) \tag{3.22}$$
$$= \frac{m+2}{m+1}\rho_{m+2}(\alpha) + \frac{\rho_{m+2}(1)}{m+1} + \rho_{m+1}(\alpha)\rho_1(\alpha) - \alpha\rho'_{m+1}(\alpha).$$

For $m = 0$ we get

$$0 = 2\rho_2(\alpha) + \rho_1^2(\alpha) - \alpha\rho_1'(\alpha) + \rho_2(1), \tag{3.23}$$

and this is just (1.14), which in turn is equivalent to Ramanujan's identity (1.5). Putting $m = 1$, 2, 3, 4 and 5 in (3.22), and noting from (3.20) that $\rho_m(1) = 0$ if $m \geq 3$ is odd, we get

$$\rho_1(\alpha)\rho_2(1) = \frac{3}{2}\rho_3(\alpha) + \rho_2(\alpha)\rho_1(\alpha) - \alpha\rho_2'(\alpha), \tag{3.24}$$

$$2\rho_2(\alpha)\rho_2(1) = \frac{4}{3}\rho_4(\alpha) + \rho_3(\alpha)\rho_1(\alpha) - \alpha\rho_3'(\alpha) + \frac{\rho_4(1)}{3}, \tag{3.25}$$

$$3\rho_3(\alpha)\rho_2(1) + \rho_1(\alpha)\rho_4(1) = \frac{5}{4}\rho_5(\alpha) + \rho_4(\alpha)\rho_1(\alpha) - \alpha\rho_4'(\alpha), \tag{3.26}$$

$$4\rho_4(\alpha)\rho_2(1)+4\rho_2(\alpha)\rho_4(1) = \frac{6}{5}\rho_6(\alpha)+\rho_5(\alpha)\rho_1(\alpha)-\alpha\rho_5'(\alpha)+\frac{\rho_6(1)}{4}, \quad (3.27)$$

and

$$5\rho_5(\alpha)\rho_2(1) + 10\rho_3(\alpha)\rho_4(1) + \rho_1(\alpha)\rho_6(1) = \frac{7}{6}\rho_7(\alpha) + \rho_6(\alpha)\rho_1(\alpha) - \alpha\rho_6'(\alpha). \quad (3.28)$$

We next put $\alpha = e^{ia}$ and define $S_{2n-1}(a)$ and $C_{2n}(a)$ for $n \geq 1$ by

$$S_{2n-1}(a) = \frac{1}{2i}\rho_{2n-1}(\alpha) \qquad \text{and} \qquad C_{2n}(a) = \frac{1}{2}\rho_{2n}(\alpha). \quad (3.29)$$

We note that

$$S_1(a) = \phi_1(a) \qquad \text{and} \qquad S_2(a) = \phi_2(a) \quad (3.30)$$

where ϕ_1 and ϕ_2 are the functions introduced earlier in (1.15). By (3.19) we have

$$S_1(a) = \frac{1}{4}\cot\frac{a}{2} + \frac{1}{2i}\sum_{j=1}^{\infty}\left(\frac{e^{ia}q^j}{1-e^{ia}q^j} - \frac{e^{-ia}q^j}{1-e^{-ia}q^j}\right) \quad (3.31)$$

and by (3.20) we have, for $n \geq 1$,

$$\begin{aligned} C_{2n}(a) &= -\frac{B_{2n}}{4n} + \frac{1}{2}\sum_{j=1}^{\infty} j^{2n-1}\left(\frac{e^{ia}q^j}{1-e^{ia}q^j} + \frac{e^{-ia}q^j}{1-e^{-ia}q^j}\right) \quad (3.32)\\ &= -\frac{B_{2n}}{4n} + \sum_{j=1}^{\infty}\sum_{k=1}^{\infty} j^{2n-1}q^{jk}\cos ka \end{aligned}$$

and

$$\begin{aligned} S_{2n+1}(a) &= \frac{1}{2i}\sum_{j=1}^{\infty} j^{2n}\left(\frac{e^{ia}q^j}{1-e^{ia}q^j} - \frac{e^{-ia}q^j}{1-e^{-ia}q^j}\right) \quad (3.33)\\ &= \sum_{j=1}^{\infty}\sum_{k=1}^{\infty} j^{2n}q^{jk}\sin ka. \end{aligned}$$

We write the formulas (3.23)–(3.28) in terms of S_{2n-1} and C_{2n}:

$$0 = C_2(a) - S_1^2(a) - \frac{1}{2}S_1'(a) + \frac{1}{2}C_2(0), \quad (3.34)$$

$$S_1(a)C_2(0) = \frac{3}{4}S_3(a) + C_2(a)S_1(a) + \frac{1}{2}C_2'(a), \quad (3.35)$$

$$2C_2(a)C_2(0) = \frac{2}{3}C_4(a) - S_3(a)S_1(a) - \frac{1}{2}S_3'(a) + \frac{1}{6}C_4(0), \quad (3.36)$$

$$3S_3(a)C_2(0) + S_1(a)C_4(0) = \frac{5}{8}S_5(a) + C_4(a)S_1(a) + \frac{1}{2}C_4'(a), \quad (3.37)$$

$$4C_4(a)C_2(0) + 4C_2(a)C_4(0) = \frac{3}{5}C_6(a) - S_5(a)S_1(a) - \frac{1}{2}S_5'(a) + \frac{1}{10}C_6(0), \quad (3.38)$$

and

$$5S_5(a)C_2(0)+10S_3(a)C_4(0)+S_1(a)C_6(0) = \frac{7}{12}S_7(a)+C_6(a)S_1(a)+\frac{1}{2}C_6'(a). \quad (3.39)$$

By (2.1) and (3.32) we have

$$C_{2n}(0) = -\frac{B_{2n}}{4n} + \sum_{j=1}^{\infty} \frac{j^{2n-1}q^j}{1-q^j} = -\frac{B_{2n}}{4n} + \phi_{0,2n-1}.$$

Therefore,

$$C_2(0) = -\frac{P}{24}. \quad (3.40)$$

Furthermore, by Ramanujan's result (2.9), for $n \geq 2$, $C_{2n}(0)$ may be expressed as a polynomial in Q and R with rational coefficients.

By (3.34) we have

$$C_2(a) = S_1^2(a) + \frac{1}{2}S_1'(a) - \frac{1}{2}C_2(0), \quad (3.41)$$

so $C_2(a)$ is a polynomial in $S_1(a)$, $S_1'(a)$ and P. Differentiating with respect to a, we obtain that $C_2'(a)$ is a polynomial in $S_1(a)$, $S_1'(a)$ and $S_1''(a)$.

Let Ω denote the set of polynomials in

$$S_1(a),\ S_1'(a),\ S_1''(a),\ P,\ Q \quad \text{and} \quad R$$

with rational coefficients. By (3.35), (3.40) and (3.41) we deduce that $S_3(a) \in \Omega$.

By (1.15), (1.28), (1.36) and (3.29) we have

$$\wp(a) = 2(\phi_2(0) - \phi_1'(a)) = 2\left(-\frac{P}{24} - S_1'(a)\right),$$

so the Weierstrass differential equation (1.40) may be expressed in the form

$$4\left(S_1'(a) + \frac{P}{24}\right)^2 + \frac{S_1'''(a)}{3} = \frac{Q}{144}. \quad (3.42)$$

Therefore, $S_1'''(a)$ belongs to Ω. By induction and repeated differentiation of (3.42) with respect to a, it follows that the higher derivatives $S_1^{(4)}(a)$,

$S_1^{(5)}(a), \ldots$, also belong to Ω. More generally, the derivative of any function that belongs to Ω will also belong to Ω.

Next, from (3.36), $C_4(a)$ belongs to Ω, and hence so do its derivatives $C_4'(a)$, $C_4''(a), \ldots$. Then, from (3.37), $S_5(a)$ and its derivatives belong to Ω.

Continuing in this way using (3.22) and by applying induction, we obtain the interesting result that for any positive integer n, the functions $S_{2n-1}(a)$ and $C_{2n}(a)$ are polynomials in

$$S_1(a),\ S_1'(a),\ S_1''(a),\ P,\ Q, \quad \text{and} \quad R$$

with rational coefficients. This is also true of their derivatives of any order with respect to a. By (3.32) and (3.33) it follows that

$$\sum_{j=1}^{\infty}\sum_{k=1}^{\infty} j^m k^n q^{mn} \cos ka, \qquad j+k \text{ odd}, \quad j,k \geq 1,$$

and

$$\sum_{j=1}^{\infty}\sum_{k=1}^{\infty} j^m k^n q^{mn} \sin ka, \qquad j+k \text{ even}, \quad j,k \geq 1,$$

are polynomials in $S_1(a)$, $S_1'(a)$, $S_1''(a)$, P, Q and R with rational coefficients. If one desires, one can eliminate R using the differential equation (1.42). This generalizes the result of Ramanujan given in Theorem 2.1, (ii).

If we substitute $\alpha = e^{ia}$, $\beta = e^{ib}$ in (3.21) and successively set $m = 0$, 1, 2, 3 and 4, we obtain

$$S_1(a)S_1(b) = -\frac{1}{2}(C_2(a) + C_2(b) + C_2(a+b)) + S_1(a+b)(S_1(a) + S_1(b)), \tag{3.43}$$

$$\begin{aligned} &S_1(a)C_2(b) + C_2(a)S_1(b) \\ &= \frac{1}{4}(S_3(a) + S_3(b) + 2S_3(a+b)) + C_2(a+b)(S_1(a) + S_1(b)), \end{aligned} \tag{3.44}$$

$$\begin{aligned} &S_1(a)S_3(b) - 2C_2(a)C_2(b) + S_3(a)S_1(b) \\ &= -\frac{1}{6}(C_4(a) + C_4(b) + 3C_4(a+b)) + S_3(a+b)(S_1(a) + S_1(b)), \end{aligned} \tag{3.45}$$

$$\begin{aligned} &S_1(a)C_4(b) + C_2(a)S_3(b) + S_3(a)C_2(b) + C_4(a)S_1(b) \\ &= \frac{1}{8}(S_5(a) + S_5(b) + 4S_5(a+b)) + C_4(a+b)(S_1(a) + S_1(b)), \end{aligned} \tag{3.46}$$

and

$$\begin{aligned} &S_1(a)S_5(b) - 4C_2(a)C_4(b) + 6S_3(a)S_3(b) - 4C_4(a)C_2(b) + S_5(a)S_1(b) \\ &= -\frac{1}{10}(C_6(a) + C_6(b) + 5C_6(a+b)) + S_5(a+b)(S_1(a) + S_1(b)). \end{aligned} \tag{3.47}$$

The identity (3.43) is equivalent to the generalized Ramanujan identity (1.9) or (1.21). The right hand sides of the identities (3.43)–(3.47) contain both the functions $C_{2m}(a+b)$ and $S_{2m\pm1}(a+b)$. In order to obtain formulas containing only one of them we apply the operator $\partial/\partial a - \partial/\partial b$. Applying the operator to (3.43) we obtain

$$S_1'(a)S_1(b) - S_1(a)S_1'(b) = -\frac{1}{2}(C_2'(a) - C_2'(b)) + S_1(a+b)(S_1'(a) - S_1'(b)).$$

We have already seen that $C_2'(a)$ (resp. $C_2'(b)$) is a polynomial in $S_1(a)$, $S_1'(a)$ and $S_1''(a)$ (resp. $S_1(b)$, $S_1'(b)$ and $S_1''(b)$). Hence we obtain that $S_1(a+b)(S_1'(a) - S_1'(b))$ is a polynomial in $S_1(a)$, $S_1'(a)$, $S_1''(a)$, $S_1(b)$, $S_1'(b)$ and $S_1''(b)$. Using this information in (3.43), it follows that $C_2(a+b)(S_1'(a) - S_1'(b))$ is a polynomial in the same variables.

If we apply the operator $\partial/\partial a - \partial/\partial b$ to (3.44) we obtain

$$\begin{aligned} &S_1'(a)C_2(b) - S_1(a)C_2'(b) + C_2'(a)S_1(b) - C_2(a)S_1'(b) \\ &= \frac{1}{4}(S_3'(a) - S_3'(b)) + C_2(a+b)(S_1'(a) - S_1'(b)). \end{aligned} \tag{3.48}$$

We have already seen that $S_1(a)$, $C_2(a)$, $S_3(a)$ and all of their derivatives can be expressed in terms of $S_1(a)$, $S_1'(a)$, $S_1''(a)$, P, Q and R. Thus, we obtain that $C_2(a+b)(S_1'(a) - S_1'(b))$ can be expressed as a polynomial with rational coefficients in

$$S_1(a),\ S_1'(a),\ S_1''(a),\ S_1(b),\ S_1'(b),\ S_1''(b),\ P,\ Q \quad \text{and} \quad R.$$

Using this information in (3.44), it follows that $S_3(a+b)(S_1'(a) - S_1'(b))$ is a polynomial in the same variables.

We can proceed successively, using (3.45), (3.46), (3.47), etc., and obtain that for any positive integer n,

$$S_{2n-1}(a+b)(S_1'(a) - S_1'(b)) \qquad \text{and} \qquad C_{2n}(a+b)(S_1'(a) - S_1'(b))$$

are all polynomials with rational coefficients in

$$S_1(a),\ S_1'(a),\ S_1''(a),\ S_1(b),\ S_1'(b),\ S_1''(b),\ P,\ Q \quad \text{and} \quad R.$$

If one desires, one can eliminate R using the differential equation (1.42).

3.3 Partitions

The identity (3.21) yields more interesting and useful information. We give here a proof of the following very beautiful identity of Ramanujan:

$$\begin{aligned} \sum_{n=0}^{\infty} \frac{q^{5n+1}}{(1-q^{5n+1})^2} - \frac{q^{5n+2}}{(1-q^{5n+2})^2} - \frac{q^{5n+3}}{(1-q^{5n+3})^2} + \frac{q^{5n+4}}{(1-q^{5n+4})^2} \\ = q\prod_{n=1}^{\infty} \frac{(1-q^{5n})^5}{(1-q^n)}. \end{aligned} \tag{3.49}$$

This identity is contained in one of Ramanujan's partially published long manuscripts [93, p. 139, (4.5)], which had been sent to Hardy after Ramanujan's death, and which came to light only after the death of G. N. Watson to whom Hardy sent the manuscript earlier.

Before giving a proof of the identity (3.49), we give an application. A partition of n is a representation of n as a sum of positive integers, in non-increasing order. The number of partitions of n is denoted by $p(n)$. For example,

$$4 = 3 + 1 = 2 + 2 = 2 + 1 + 1 = 1 + 1 + 1 + 1,$$

so the integer 4 has 5 partitions, and $p(4) = 5$. By convention, $p(0)$ is defined to be 1. The generating function for $p(n)$ is given by

$$\sum_{n=0}^{\infty} p(n)q^n = \prod_{n=1}^{\infty} \frac{1}{1-q^n}.$$

From the identity (3.49), Ramanujan deduced in a very simple way another of his famous identities:

$$\sum_{n=0}^{\infty} p(5n+4)q^n = 5\prod_{n=1}^{\infty} \frac{(1-q^{5n})^5}{(1-q^n)^6}. \tag{3.50}$$

Hardy [94, p. xxxv] wrote "*... if I had to select one formula from all Ramanujan's work, I would agree with Major MacMahon in selecting* (3.50)."

Ramanujan's deduction of (3.50) from (3.49) is given in [93, pp. 139–140]; see also the paper by K. G. Ramanathan [89].

We shall now give a proof of Ramanujan's identity (3.49), using the identity (3.48). By (3.31), (3.32) and (3.33), we have the expansions

$$S_1(a) = \frac{1}{4}\cot\frac{a}{2} + \sum_{n=1}^{\infty}\sum_{m=1}^{\infty} q^{mn}\sin ma, \tag{3.51}$$

$$S_1'(a) = -\frac{1}{8}\csc^2\frac{a}{2} + \sum_{n=1}^{\infty}\sum_{m=1}^{\infty} mq^{mn}\cos ma, \tag{3.52}$$

$$C_2(a) = -\frac{1}{24} + \sum_{n=1}^{\infty}\sum_{m=1}^{\infty} nq^{mn}\cos ma, \tag{3.53}$$

$$C_2'(a) = -\sum_{n=1}^{\infty}\sum_{m=1}^{\infty} mnq^{mn}\sin ma, \tag{3.54}$$

and

$$S_3'(a) = \sum_{n=1}^{\infty}\sum_{m=1}^{\infty} mn^2 q^{mn} \cos ma = q\frac{\partial}{\partial q} C_2(a). \tag{3.55}$$

We note that

$$S_1\left(\frac{6\pi}{5}\right) = -S_1\left(\frac{4\pi}{5}\right), \quad S_1'\left(\frac{6\pi}{5}\right) = S_1'\left(\frac{4\pi}{5}\right), \quad C_2\left(\frac{6\pi}{5}\right) = C_2\left(\frac{4\pi}{5}\right),$$

$$C_2\left(\frac{8\pi}{5}\right) = C_2\left(\frac{2\pi}{5}\right), \quad C_2'\left(\frac{6\pi}{5}\right) = -C_2'\left(\frac{4\pi}{5}\right), \quad S_3'\left(\frac{6\pi}{5}\right) = S_3'\left(\frac{4\pi}{5}\right).$$

We now put $(a, b) = (2\pi/5, 4\pi/5)$ in (3.48), then put $(a, b) = (2\pi/5, 6\pi/5)$ in (3.48) and add, to get

$$\begin{aligned} &2\left(S_1'\left(\frac{2\pi}{5}\right)C_2\left(\frac{4\pi}{5}\right) - C_2\left(\frac{2\pi}{5}\right)S_1'\left(\frac{4\pi}{5}\right)\right) \\ &= \frac{1}{2}\left(S_3'\left(\frac{2\pi}{5}\right) - S_3'\left(\frac{4\pi}{5}\right)\right) \\ &\quad + \left(C_2\left(\frac{2\pi}{5}\right) + C_2\left(\frac{4\pi}{5}\right)\right)\left(S_1'\left(\frac{2\pi}{5}\right) - S_1'\left(\frac{4\pi}{5}\right)\right). \end{aligned} \tag{3.56}$$

Using (3.51)–(3.55), together with the evaluations

$$\cos\left(\frac{2m\pi}{5}\right) = \begin{cases} \dfrac{-1+\sqrt{5}}{4} & \text{if } m \equiv 1 \text{ or } 4 \pmod 5, \\ \dfrac{-1-\sqrt{5}}{4} & \text{if } m \equiv 2 \text{ or } 3 \pmod 5, \\ 1 & \text{if } m \equiv 0 \pmod 5 \end{cases}$$

and

$$\csc^2\left(\frac{\pi}{5}\right) = 2\left(1 + \frac{\sqrt{5}}{5}\right), \qquad \csc^2\left(\frac{2\pi}{5}\right) = 2\left(1 - \frac{\sqrt{5}}{5}\right),$$

we find that

$$C_2\left(\frac{2\pi}{5}\right) = A + B\sqrt{5}, \qquad C_2\left(\frac{4\pi}{5}\right) = A - B\sqrt{5},$$

$$S_1'\left(\frac{2\pi}{5}\right) = C + D\sqrt{5}, \qquad S_1'\left(\frac{4\pi}{5}\right) = C - D\sqrt{5},$$

$$S_3'\left(\frac{2\pi}{5}\right) = q\frac{d}{dq}(A + B\sqrt{5}), \qquad S_3'\left(\frac{4\pi}{5}\right) = q\frac{d}{dq}(A - B\sqrt{5})$$

where

$$A = -\frac{1}{24} - \frac{1}{4}\sum_{n=1}^{\infty} \frac{q^n}{(1-q^n)^2} + \frac{5}{4}\sum_{n=1}^{\infty} \frac{q^{5n}}{(1-q^{5n})^2},$$

$$B = \frac{1}{4}\sum_{n=0}^{\infty}\left(\frac{q^{5n+1}}{(1-q^{5n+1})^2} - \frac{q^{5n+2}}{(1-q^{5n+2})^2} - \frac{q^{5n+3}}{(1-q^{5n+3})^2} + \frac{q^{5n+4}}{(1-q^{5n+4})^2}\right),$$

$$C = -\frac{1}{4} - \frac{1}{4}\sum_{n=1}^{\infty} \frac{nq^n}{1-q^n} + \frac{25}{4}\sum_{n=1}^{\infty} \frac{nq^{5n}}{1-q^{5n}},$$

$$D = -\frac{1}{20} + \frac{1}{4}\sum_{n=0}^{\infty}\left(\frac{(5n+1)q^{5n+1}}{1-q^{5n+1}} - \frac{(5n+2)q^{5n+2}}{1-q^{5n+2}} - \frac{(5n+3)q^{5n+3}}{1-q^{5n+3}} + \frac{(5n+4)q^{5n+4}}{1-q^{5n+4}}\right).$$

The identity (3.56), expressed in terms of A, B, C and D, is

$$2\left((C+D\sqrt{5})(A-B\sqrt{5}) - (A+B\sqrt{5})(C-D\sqrt{5})\right) = \frac{1}{2}\left(q\frac{d}{dq}(A+B\sqrt{5}) - q\frac{d}{dq}(A-B\sqrt{5})\right) + 4AD\sqrt{5},$$

which simplifies to

$$-4BC = q\frac{dB}{dq}.$$

Thus,

$$\frac{d}{dq}\log B = -\frac{4}{q}C = \frac{1}{q} + \sum_{n=1}^{\infty}\frac{nq^{n-1}}{1-q^n} - 25\sum_{n=1}^{\infty}\frac{nq^{5n-1}}{1-q^{5n}}.$$

Now integrate both sides and exponentiate to get

$$B = kq\prod_{n=1}^{\infty}\frac{(1-q^{5n})^5}{(1-q^n)},$$

for some constant k. Comparing coefficients of q on both sides implies $k = 1/4$. This proves Ramanujan's identity (3.49).

3.4 The hypergeometric function ${}_2F_1(1/2, 1/2; 1; x)$: first method

Ramanujan introduces his invariants z and x in his own interesting and unique manner in Chapter 17 of his second notebook [92]. We shall now give a method for deriving the classical hypergeometric differential equation satisfied by z, using Ramanujan's differential equations.

Let us suppose in a neighborhood of $q=0$ that x and z are algebraically dependent on Q and R according to the formulas[1]

$$Q = 1 + 240\sum_{n=1}^{\infty}\frac{n^3q^n}{1-q^n} = z^4(1+14x+x^2) \tag{3.57}$$

and

$$R = 1 - 504\sum_{n=1}^{\infty}\frac{n^5q^n}{1-q^n} = z^6(1+x)(1-34x+x^2). \tag{3.58}$$

When $q=0$ we have $Q=R=1$, and the only solutions of the system

$$z^4(1+14x+x^2)=1, \qquad z^6(1+x)(1-34x+x^2)=1$$

are given by

$$(x,z)=(0,\pm 1),\ (1,\pm 1/2) \quad \text{and} \quad (1,\pm i/2).$$

We consider the solution that satisfies $(x,z)=(0,1)$ when $q=0$. Let f_1 and f_2 be defined by

$$f_1(q,x,z) = z^4(1+14x+x^2) - Q(q),$$

and

$$f_2(q,x,z) = z^6(1+x)(1-34x+x^2) - R(q),$$

and note that $f_1(0,0,1)=f_2(0,0,1)=0$. We find that

$$\det\begin{pmatrix} \dfrac{\partial f_1}{\partial x} & \dfrac{\partial f_1}{\partial z} \\ \\ \dfrac{\partial f_2}{\partial x} & \dfrac{\partial f_2}{\partial z}\end{pmatrix} = 216z^9(1-x)^3$$

and this determinant is not zero when $(x,z)=(0,1)$. Therefore by the implicit function theorem, there are unique analytic functions $x=x(q)$ and $z=z(q)$ that satisfy the equations (3.57) and (3.58) and the initial conditions $x(0)=0$ and $z(0)=1$. For future reference, we compute the first few terms in the expansions of x and z in powers of q. We write

$$x = a_1q + a_2q^2 + \cdots \qquad \text{and} \qquad z = 1 + b_1q + b_2q^2 + \cdots,$$

[1] The formulas (3.57) and (3.58) are introduced here without any explanation. A possible motivation is given in the notes at the end of the chapter.

where a_1, a_2, b_1 and b_2 are coefficients that we will determine. Expanding (3.57) and (3.58) as far as the first power of q we get

$$1 + 240q + O(q^2) = 1 + (14a_1 + 4b_1)q + O(q^2)$$

and

$$1 - 504q + O(q^2) = 1 + (-33a_1 + 6b_1)q + O(q^2).$$

Equating the coefficients of q and solving the resulting system of equations, we find that $a_1 = 16$ and $b_1 = 4$. Making use of these values and expanding (3.57) and (3.58) as far as the second power of q we get

$$1 + 240q + 2160q^2 + O(q^3) = 1 + 240q + (3936 + 14a_2 + 4b_2)q^2 + O(q^3)$$

and

$$1 - 504q - 16632q^2 + O(q^3) = 1 - 504q + (-20880 - 33a_2 + 6b_2)q^2 + O(q^3).$$

Equating coefficients of q^2 we find that $a_2 = -128$ and $b_2 = 4$. Hence,

$$x = 16q - 128q^2 + O(q^3) \qquad \text{and} \qquad z = 1 + 4q + 4q^2 + O(q^3). \tag{3.59}$$

Furthermore, since

$$\det \begin{pmatrix} \dfrac{\partial f_1}{\partial q} & \dfrac{\partial f_1}{\partial z} \\ \dfrac{\partial f_2}{\partial q} & \dfrac{\partial f_2}{\partial z} \end{pmatrix} \Bigg|_{(q,x,z)=(0,0,1)} = \det \begin{pmatrix} -240 & 4 \\ 504 & 6 \end{pmatrix} \neq 0,$$

it follows that q and z may be regarded as functions of x in a neighborhood of $x = 0$; we will find explicit formulas for these functions.

We denote differentiation with respect to $\log q$ by $'$. That is,

$$u' = \frac{du}{d\log q} = q\frac{du}{dq}.$$

By a straightforward calculation using (3.57) and (3.58), we find that

$$Q^3 - R^2 = 108\, z^{12} x(1-x)^4.$$

Differentiating this logarithmically using (2.14), we get

$$P = 12\frac{z'}{z} + \left(\frac{1}{x} - \frac{4}{1-x}\right)x'. \tag{3.60}$$

Applying logarithmic differentiation to (3.57) we get

$$\frac{Q'}{Q} = \frac{4z'}{z} + \frac{(14+2x)x'}{1+14x+x^2}, \tag{3.61}$$

and by Ramanujan's differential equations (2.13), we have

$$\frac{3Q'}{Q} = P - \frac{R}{Q}. \tag{3.62}$$

Now substitute (3.60) and (3.61) into (3.62) to eliminate the P and Q'/Q terms, respectively, to get

$$12\frac{z'}{z} + \frac{3(14+2x)}{1+14x+x^2}x' = 12\frac{z'}{z} + \left(\frac{1}{x} - \frac{4}{1-x}\right)x' - \frac{R}{Q}.$$

This simplifies to

$$x' = \frac{R}{Q}\,\frac{x(1-x)(1+14x+x^2)}{(1+x)(1-34x+x^2)}.$$

Now substituting for Q and R from (3.57) and (3.58), this simplifies further to

$$x' = q\frac{dx}{dq} = z^2x(1-x). \tag{3.63}$$

We substitute (3.63) into (3.60) to eliminate the x' term, and obtain

$$z' = q\frac{dz}{dq} = \frac{1}{12}\left(Pz - (1-5x)z^3\right). \tag{3.64}$$

We now prove the classical hypergeometric differential equation satisfied by z with respect to x. By (3.63) and (3.64), we have

$$\frac{dz}{dx} = \frac{z'}{x'} = \frac{P-(1-5x)z^2}{12x(1-x)z}$$

and so

$$12x(1-x)\frac{dz}{dx} = \frac{P}{z} - (1-5x)z. \tag{3.65}$$

Differentiate with respect to x to get

$$\begin{aligned} 12\frac{d}{dx}\left(x(1-x)\frac{dz}{dx}\right) &= \frac{d}{dx}\left(\frac{P}{z} - (1-5x)z\right) \\ &= q\frac{d}{dq}\left(\frac{P}{z} - (1-5x)z\right)\bigg/\left(q\frac{dx}{dq}\right). \end{aligned}$$

Now use (2.13), (3.63) and (3.64) to calculate the derivatives on the right hand side, and simplify. The result is the classical hypergeometric differential equation

$$\frac{d}{dx}\left(x(1-x)\frac{dz}{dx}\right) = \frac{z}{4}. \tag{3.66}$$

The hypergeometric function ${}_2F_1$ is defined by

$${}_2F_1(a,b;c;x) = 1 + \sum_{n=1}^{\infty} \frac{(a)_n (b)_n}{(c)_n n!} x^n,$$

where a, b and c are complex numbers, $|x| < 1$ and

$$(a)_n = a(a+1)(a+2)\cdots(a+n-1),$$

for each positive integer n, and $(a)_0 = 1$. The general solution of (3.66) is

$$z = A\ {}_2F_1(1/2,1/2;1;x) + B\ {}_2F_1(1/2,1/2;1;1-x) \tag{3.67}$$

for arbitrary constants A and B. By the theorem of comparison for two divergent series [36, p. 149] and Wallis' product formula for π [36, p. 213], we obtain

$$\lim_{x \to 0^+} \frac{{}_2F_1(1/2,1/2;1;1-x)}{-\log x} = \lim_{n\to\infty} \left(\frac{(1/2)_n}{n!}\right)^2 \Big/ \left(\frac{1}{n}\right) = \frac{1}{\pi}. \tag{3.68}$$

Since $z = 1$ when $x = 0$, we deduce from (3.67) and (3.68) that $A = 1$ and $B = 0$, and hence

$$z = {}_2F_1(1/2,1/2;1;x). \tag{3.69}$$

We shall now verify that $z \log q$ is also a solution of of the hypergeometric equation (3.66). Put $\log q = t$. Then from (3.63) we have

$$\frac{dx}{dt} = z^2 x(1-x)$$

or

$$x(1-x)\frac{dt}{dx} = \frac{1}{z^2}. \tag{3.70}$$

Next,

$$\begin{aligned} x(1-x)\frac{d}{dx}(tz) &= zx(1-x)\frac{dt}{dx} + tx(1-x)\frac{dz}{dx} \\ &= \frac{1}{z} + tx(1-x)\frac{dz}{dx}. \end{aligned}$$

Differentiating again and using (3.66) and (3.70) gives

$$\begin{aligned} \frac{d}{dx}\left(x(1-x)\frac{d}{dx}(tz)\right) &= -\frac{1}{z^2}\frac{dz}{dx} + x(1-x)\frac{dt}{dx}\frac{dz}{dx} + t\frac{d}{dx}\left(x(1-x)\frac{dz}{dx}\right) \\ &= -\frac{1}{z^2}\frac{dz}{dx} + \frac{1}{z^2}\frac{dz}{dx} + \frac{tz}{4} \\ &= \frac{tz}{4}. \end{aligned}$$

Hence $tz = z \log q$ is also a solution of the hypergeometric differential equation, (3.66) and therefore it may be expressed in terms of the general solution (3.67). This gives

$$tz = {}_2F_1(1/2,1/2;1;x)\log q = C\,{}_2F_1(1/2,1/2;1;x) + D\,{}_2F_1(1/2,1/2;1;1-x)$$

or

$$\log q = C + D\frac{{}_2F_1(1/2,1/2;1;1-x)}{{}_2F_1(1/2,1/2;1;x)}, \tag{3.71}$$

for some constants C and D. We divide both sides by $\log x$ and take the limit as $x \to 0^+$ to get

$$\lim_{x\to 0^+} \frac{\log q}{\log x} = D \lim_{x\to 0^+} \frac{{}_2F_1(1/2,1/2;1;1-x)}{\log x}. \tag{3.72}$$

The limit on the left hand side of (3.72) may be determined using L'Hôpital's rule and (3.63):

$$\lim_{x\to 0^+} \frac{\log q}{\log x} = \lim_{x\to 0^+} \left(\frac{1}{q}\frac{dq}{dx}\right) \Big/ \left(\frac{1}{x}\right) = \lim_{x\to 0^+} \frac{1}{z^2(1-x)} = 1,$$

and the limit on the right hand side of (3.72) follows from (3.68). Therefore, $D = -\pi$ and (3.71) may be written in the form

$$q = E \exp\left(-\pi \frac{{}_2F_1(1/2,1/2;1;1-x)}{{}_2F_1(1/2,1/2;1;x)}\right) \tag{3.73}$$

for some constant E. In order to determine E we divide by x and take the limit as $q \to 0^+$. By (3.59) we have $\lim_{q\to 0^+} x = 0$ and moreover, $\lim_{q\to 0^+} q/x = 1/16$. On the other hand, using the result (e.g., see [111, p. 522])

$$\lim_{x\to 0^+} (\pi\,{}_2F_1(1/2,1/2;1;1-x) + \log x) = \log 16, \tag{3.74}$$

we deduce from (3.73) that

$$\lim_{q\to 0^+} \frac{q}{x} = \lim_{x\to 0^+} E \exp\left(-\pi \frac{{}_2F_1(1/2,1/2;1;1-x)}{{}_2F_1(1/2,1/2;1;x)} - \log x\right) = \frac{E}{16}.$$

It follows that $E = 1$, and we have proved:

$$q = \exp\left(-\pi \frac{{}_2F_1(1/2,1/2;1;1-x)}{{}_2F_1(1/2,1/2;1;x)}\right). \tag{3.75}$$

3.5 Notes

The formulas (3.4) and (3.14) were given by L. Kronecker in lectures in July 1876, and then in a paper published in 1881 [78]. In fact, Kronecker gives the identity

$$\frac{\sum(-1)^{(\mu-1)/2}\mu q^{\mu^2/4}\sum(-1)^{(\nu-1)/2}q^{\nu^2/4}(x^\nu y^\nu - x^{-\nu}y^{-\nu})}{\sum(-q)^{m^2}x^{2m}\sum(-q)^{n^2}y^{2n}}$$
$$= \sum\sum q^{\mu\nu/2}(x^\mu y^\nu - x^{-\mu}y^{-\nu}),$$

where the sums are over positive odd integers μ and ν, and all integers m and n. Kronecker's form can be manipulated into the form (3.14) by converting each of the four sums in the quotient into infinite products via Jacobi's triple product identity and making a change of variable. The formulas (3.1), (3.4), (3.7) and (3.14) appear in Jordan's Cours d'Analyse [75, p 507–511]. It is likely that for these reasons Venkatachaliengar refers to the function $f(\alpha, t)$ as the Jordan-Kronecker function.

Kroencker's paper has a reference to the 1850 paper of Jacobi "Sur la rotation d'un corps", [71]. If eq. (3) in [71, p. 297] is multiplied by $i = \sqrt{-1}$ and the result subtracted from eq. (4), *ibid.*, the result eventually simplifies to the identity (3.14). Thus, although (3.14) is implicit in Jacobi's work, it would be a bit of a stretch to attribute the identity to him. However, Jacobi's work is noteworthy because the genesis of the identity may be seen there, and because the identity arises in a physical application.

Kronecker's analysis has been examined and simplified by Weil [108, pp. 70–71]. Both Kronecker and Weil start with the product in (3.14), expand as a Laurent series in powers of t, and use Cauchy's theorem to compute the coefficients.

The fundamental multiplicative identity (3.17) should be regarded as classical. It has been rediscovered and reproved many times. For, replacing α and β with x and y, respectively, and then dividing by $f(xy, t)$ gives

$$\begin{aligned}\frac{f(x,t)f(y,t)}{f(xy,t)} &= t\frac{\partial}{\partial t}\log f(xy,t) + \rho_1(x) + \rho_1(y)\\ &= \rho_1(x) + \rho_1(y) + \rho_1(t) - \rho_1(xyt),\end{aligned}$$

and this can be shown to be equivalent to the Weierstrassian identity

$$\frac{\sigma(\alpha+\beta)\sigma(\alpha-\beta)\sigma(2\gamma)}{\sigma(\alpha+\gamma)\sigma(\alpha-\gamma)\sigma(\beta+\gamma)\sigma(\beta-\gamma)} = \zeta(\alpha+\gamma)-\zeta(\alpha-\gamma)+\zeta(\beta-\gamma)-\zeta(\beta+\gamma)$$

given by G.-H. Halphen [60, p. 187]. Further historical background for the identity (3.17) has been given in [4, p. 60]. For generalizations, see the work by S. H. Chan [42].

Ramanujan's identity (3.49) appears in his manuscript on the partition and tau functions in the lost notebook [93, pp. 139–140]. Ramanujan does not give a proof of the identity (3.49) but instead says *"It appears that ..."*, and uses (3.49) to deduce (3.50). The identity (3.49) appears again in the lost notebook [93] on pp. 354 and 357 in the equations labelled (1.52) and (3.82), respectively. For references to proofs of (3.49) and (3.50), see Berndt's commentary [94, pp. 372–373]. A direct and simple proof of (3.50) that uses only the Jacobi triple product identity has been given by M. D. Hirschhorn [64]. The method of proof of (3.49) given in Sec. 3.3 has been greatly extended in forthcoming work by T. Huber [68].

The material in Sec. 3.4 is interesting and original. Some of the ideas have been utilized by H. H. Chan and Y. L. Ong [41] in establishing a septic theory. However, the main drawback of the method used in Sec. 3.4 is that the formulas (3.57) and (3.58) are unmotivated and have to be known in advance. It is not the method Ramanujan used. Ramanujan in fact outlined his approach in the first few entries of Chapter 17 in his second notebook [92], and concerning this Berndt made the observation [14, p. 99]: *"Proofs in the latter half of the second notebook are very rare indeed"*. The full details of Ramanujan's method have been worked out by Berndt [14, pp. 87–102].

The theory of theta functions may be used to motivate the formulas (3.57) and (3.58) as follows. Ramanujan's theta functions $\varphi(q)$ and $\psi(q)$ are defined by

$$\varphi(q)=\sum_{n=-\infty}^{\infty} q^{n^2} \qquad \text{and} \qquad \psi(q)=\sum_{n=0}^{\infty} q^{n(n+1)/2}.$$

Jacobi's sum of eight squares and sum of eight triangular numbers formulas may be stated as

$$\varphi^8(-q)=\frac{1}{15}\left(16Q(q^2)-Q(q)\right)$$

and

$$q\psi^8(q)=\frac{1}{240}\left(Q(q)-Q(q^2)\right).$$

Many proofs of these formulas have been given, for example, see the paper by Cooper and H. Y. Lam [49] for proofs that use the fundamental multiplicative identity (3.17), and see [16, Sec. 3.8] for other references and more information. It follows that

$$Q(q) = \varphi^8(-q) + 256q\psi^8(q). \tag{3.76}$$

We quote two simple results from the theory of theta functions:

$$\varphi^4(q) = \varphi^4(-q) + 16q\psi^4(q^2) \tag{3.77}$$

and

$$\left(\varphi^4(q) + 16q\psi^4(q^2)\right)^2 = \varphi^8(-q) + 64q\psi^8(q). \tag{3.78}$$

For proofs, see [16, p. 72] and [15, p. 151], respectively. In fact, we will prove (3.77) in the next chapter—see (4.40). On using (3.77) and (3.78) in (3.76), we obtain

$$\begin{aligned}
Q(q) &= 4\left(\varphi^8(-q) + 64q\psi^8(q)\right) - 3\varphi^8(q) \\
&= 4\left(\varphi^4(q) + 16q\psi^4(q^2)\right)^2 - 3\left(\varphi^4(q) - 16q\psi^4(q^2)\right)^2 \\
&= \varphi^8(q) + 224q\varphi^4(q)\psi^4(q^2) + 256q^2\psi^8(q^2) \\
&= z^4(1 + 14x + x^2),
\end{aligned}$$

where

$$z = \varphi^2(q) \qquad \text{and} \qquad x = 16q\frac{\psi^4(q^2)}{\varphi^4(q)}. \tag{3.79}$$

Next, by (2.15), (2.43)–(2.45), (3.77) and (3.79), we have

$$\begin{aligned}
Q^3(q) - R^2(q) &= 1728q\prod_{j=1}^{\infty}(1 - q^j)^{24} \\
&= 1728q\varphi^4(q)\varphi^{16}(-q)\psi^4(q^2) \\
&= 1728q\varphi^4(q)\left(\varphi^4(q) - 16q\psi^4(q^2)\right)^4\psi^4(q^2) \\
&= 108z^{12}x(1 - x)^4,
\end{aligned}$$

and so

$$\begin{aligned}
R^2(q) &= Q^3(q) - 108z^{12}x(1 - x)^4 \\
&= z^{12}\left\{(1 + 14x + x^2)^3 - 108x(1 - x)^4\right\} \\
&= z^{12}(1 + x)^2(1 - 34x + x^2)^2.
\end{aligned}$$

On taking square roots, and using the fact that both sides are 1 when $q = 0$, we obtain (3.58).

The theorem of comparison for two divergent series, used to obtain (3.68), states:

Suppose that $\sum a_n$ and $\sum b_n$ are both divergent, but that

$$f(x) = \sum_{n=1}^{\infty} a_n x^n \qquad \text{and} \qquad g(x) = \sum_{n=0}^{\infty} b_n x^n$$

are absolutely convergent for $|x| < 1$. If $a_n/b_n \to \ell$ as $n \to \infty$, then

$$\lim_{x\to 1^-} \frac{f(x)}{g(x)} = \ell = \lim_{n\to\infty} \frac{a_n}{b_n}.$$

For a proof, see [36, p. 149].

Wallis' product formula for π, used in the second part of (3.68), is a special case of the infinite product expansion for the sine function. More generally, we have:

$$\begin{aligned}\lim_{n\to\infty} \left(\frac{(a)_n (1-a)_n}{n!}^2 \right) \times n &= a \lim_{n\to\infty} \frac{(a+1)_{n-1}}{(n-1)!} \frac{(1-a)_n}{n!} \qquad (3.80)\\ &= a \lim_{n\to\infty} \frac{n}{a+n} \prod_{j=1}^{n} \left(1+\frac{a}{j}\right)\left(1-\frac{a}{j}\right)\\ &= a \prod_{j=1}^{\infty} \left(1-\frac{a^2}{j^2}\right)\\ &= \frac{\sin \pi a}{\pi},\end{aligned}$$

where the last step follows by the infinite product formula for the sine function [36, p. 213], [102]. This result will be used again in Chapter 5. The Wallis product formula is the special case $a = 1/2$.

For simple and direct proofs of (3.74), see [29, p. 11], [34, p. 21] or [111, p. 522]. A more general result which holds for $0 < a < 1$ is

$$\lim_{x\to 0^+} \left(\frac{\pi}{\sin \pi a} \, {}_2F_1(a, 1-a; 1; 1-x) + \log x \right) = 2\Psi(1) - \Psi(a) - \Psi(1-a), \qquad (3.81)$$

where Ψ is the digamma function defined by

$$\Psi(x) = \frac{d}{dx} \log \Gamma(x).$$

Values of the digamma function for rational values of the argument can be computed using the formula [3, p. 13]:

$$\Psi\left(\frac{p}{q}\right) = -\gamma - \frac{\pi}{2} \cot \frac{\pi p}{q} - \log q + 2 \sum \cos \frac{2\pi n p}{q} \log \left(2 \sin \frac{\pi n}{q}\right). \qquad (3.82)$$

Here $0 < p < q$; γ is Euler's constant; the sum is over integers n that satisfy $1 \le n \le q/2$; and, when q is even, the term with index $n = q/2$ is divided by 2. Certain values of the digamma function that can be computed from (3.82) will be needed in Chapter 5, and these are recorded in Table 3.1.

Table 3.1 Values of the digamma function.

a	$\Psi(a)+\gamma$	$\Psi(1-a)+\gamma$
1	0	∞
$\frac{1}{2}$	$-2\log 2$	$-2\log 2$
$\frac{1}{3}$	$-\frac{3}{2}\log 3-\frac{\pi\sqrt{3}}{6}$	$-\frac{3}{2}\log 3+\frac{\pi\sqrt{3}}{6}$
$\frac{1}{4}$	$-3\log 2-\frac{\pi}{2}$	$-3\log 2+\frac{\pi}{2}$
$\frac{1}{6}$	$-2\log 2-\frac{3}{2}\log 3-\frac{\pi\sqrt{3}}{2}$	$-2\log 2-\frac{3}{2}\log 3+\frac{\pi\sqrt{3}}{2}$

Perhaps the easiest way to prove (3.81) is to consider the even more general result, recorded by Ramanujan in his second notebook [92, Chapter 11, Entry 26] and proved by Berndt [12, p. 78]:

$$\frac{\Gamma(a)\Gamma(b)}{\Gamma(a+b)}{}_2F_1(a,b;a+b;1-x) \tag{3.83}$$
$$=\sum_{n=0}^{\infty}\frac{(a)_n(b)_n}{n!^2}x^n\left(2\Psi(1+n)-\Psi(a+n)-\Psi(b+n)-\log x\right).$$

This may be proved as follows. By the theory of linear differential equations applied to the hypergeometric differential equation

$$x(1-x)\frac{d^2y}{dx^2}+(c-(a+b+1)x)\frac{dy}{dx}-aby=0,$$

it can be shown that [3, p. 78]

$${}_2F_1(a,b;a+b+1-c;1-x)$$
$$=A\,{}_2F_1(a,b;c;x)+B\,x^{1-c}{}_2F_1(1+a-c,1+b-c;2-c;x),$$

where

$$A=\frac{\Gamma(a+b+1-c)\Gamma(1-c)}{\Gamma(a+1-c)\Gamma(b+1-c)} \quad\text{and}\quad B=\frac{\Gamma(c-1)\Gamma(a+b+1-c)}{\Gamma(a)\Gamma(b)}.$$

We rewrite this as

$$\frac{\Gamma(a)\Gamma(b)\Gamma(a+1-c)\Gamma(b+1-c)}{\Gamma(c)\Gamma(2-c)\Gamma(a+b+1-c)}{}_2F_1(a,b;a+b+1-c;1-x)$$
$$=\frac{1}{1-c}\left(\sum_{n=0}^{\infty}\frac{\Gamma(a+n)\Gamma(b+n)}{\Gamma(c+n)\Gamma(1+n)}x^n\right.$$
$$\left.-\sum_{n=0}^{\infty}\frac{\Gamma(1+a-c+n)\Gamma(1+b-c+n)}{\Gamma(2-c+n)\Gamma(1+n)}x^{n+1-c}\right)$$

and take the limit as $c\to 1$. The result simplifies to (3.83).

The formula (3.75) was known to Jacobi [70].

Chapter 4

The Weierstrassian Invariants

4.1 Halphen's differential equations

In this chapter, we consider the Weierstrassian invariants e_1, e_2 and e_3. These are roots of the relevant cubic

$$4\wp^3 - g_2\wp - g_3 = 0.$$

In the classical treatment, there may be ambiguity in the definition of the e_r. We shall define them in a unique manner. We will obtain their q-series expansions as well as infinite products for their differences. We will also determine their imaginary and quadratic transformations.

We first of all show how the Darboux-Halphen differential equations may be solved in terms of the transcendentals of elliptic function theory by employing the generalized Ramanujan identity (1.13). This is given only by Halphen in his treatise on elliptic functions and is not treated in other books on elliptic functions. It may be mentioned that the Ramanujan function $\phi_1(\theta)$ defined by (1.15) occurs in Halphen's treatise in connection with a mechanical problem, and Halphen deduced the well-known Puiseux-Halphen inequality of the motion of the spherical pendulum.

Let τ be a complex number with positive imaginary part, and let h and q be defined by

$$h = \exp(\pi i\tau) \qquad \text{and} \qquad q = \exp(2\pi i\tau) \tag{4.1}$$

so that $h^2 = q$.

Halphen's differential equations are the simultaneous equations

$$h\frac{d}{dh}(u_1+u_2) = u_1u_2, \qquad h\frac{d}{dh}(u_1+u_3) = u_1u_3, \qquad h\frac{d}{dh}(u_2+u_3) = u_2u_3, \tag{4.2}$$

where u_1, u_2 and u_3 are analytic functions of h. G. Darboux [50, p. 149] encountered these equations in connection with a problem in mechanics and also in differential geometry and asked Halphen to find, if possible, an explicit solution. One was given in the following form by Halphen [60, p. 330]:

$$u_1 = u_1(h) = 1 - 8\sum_{n=1}^{\infty} \frac{(-1)^n n h^{2n}}{1-h^{2n}} = 1 + 8\sum_{m=1}^{\infty} \frac{h^{2m}}{(1+h^{2m})^2}, \tag{4.3}$$

$$u_2 = u_2(h) = -8\sum_{n=1}^{\infty} \frac{n h^{n}}{1-h^{2n}} = -8\sum_{m=1}^{\infty} \frac{h^{2m-1}}{(1-h^{2m-1})^2}, \tag{4.4}$$

$$u_3 = u_3(h) = -8\sum_{n=1}^{\infty} \frac{(-1)^n n h^{n}}{1-h^{2n}} = 8\sum_{m=1}^{\infty} \frac{h^{2m-1}}{(1+h^{2m-1})^2}. \tag{4.5}$$

We now verify that the functions (4.3)–(4.5) satisfy the Halphen differential equations (4.2). Differentiating the generalized Ramanujan identity (1.21) twice with respect to a gives

$$\phi_1''(a)\phi_1(b) - \phi_1''(a)\phi_1(a+b) - 2\phi_1'(a)\phi_1'(a+b) - \phi_1(a)\phi_1''(a+b) - \phi_1(b)\phi_1''(a+b) \tag{4.6}$$

$$= -\frac{1}{2}\left[\phi_2''(a) + \phi_2''(a+b)\right].$$

By (1.18)–(1.20), ϕ_1'' is an odd elliptic function with periods 2π and $2\pi\tau$, and it follows that

$$\phi_1''(\pi) = \phi_1''(\pi\tau) = \phi_1''(\pi+\pi\tau) = 0.$$

In turn, substitute $(a,b) = (\pi, \pi\tau - \pi)$, $(a,b) = (\pi, \pi\tau)$ and $(a,b) = (\pi\tau, \pi)$ into (4.6) to obtain

$$\phi_1'(\pi)\phi_1'(\pi\tau) = \frac{1}{4}(\phi_2''(\pi) + \phi_2''(\pi\tau)) \tag{4.7}$$

$$\phi_1'(\pi)\phi_1'(\pi+\pi\tau) = \frac{1}{4}(\phi_2''(\pi) + \phi_2''(\pi+\pi\tau)) \tag{4.8}$$

$$\phi_1'(\pi\tau)\phi_1'(\pi+\pi\tau) = \frac{1}{4}(\phi_2''(\pi\tau) + \phi_2''(\pi+\pi\tau)). \tag{4.9}$$

From (1.16) and (1.17) we find that

$$\phi_2''(\theta) = -\sum_{n=1}^{\infty} \frac{n^2 q^n}{(1-q^n)^2}\cos n\theta = -\frac{\partial}{\partial q}\left(\sum_{n=1}^{\infty} \frac{n q^n}{1-q^n}\cos n\theta\right)$$

$$= -q\frac{\partial}{\partial q}\phi_1'(\theta) = -\frac{h}{2}\frac{\partial}{\partial h}\phi_1'(\theta),$$

where the primes denote differentiation with respect to θ. Therefore, from (4.7)–(4.9) we obtain

$$\phi_1'(\pi)\phi_1'(\pi\tau) = -\frac{h}{8}\frac{d}{dh}(\phi_1'(\pi) + \phi_1'(\pi\tau)), \tag{4.10}$$

$$\phi_1'(\pi)\phi_1'(\pi+\pi\tau) = -\frac{h}{8}\frac{d}{dh}(\phi_1'(\pi) + \phi_1'(\pi+\pi\tau)), \tag{4.11}$$

$$\phi_1'(\pi\tau)\phi_1'(\pi+\pi\tau) = -\frac{h}{8}\frac{d}{dh}(\phi_1'(\pi\tau) + \phi_1'(\pi+\pi\tau)). \tag{4.12}$$

Next, from (1.16) and writing $q = h^2$, we have

$$\phi_1'(\theta) = -\frac{1}{8}\csc^2\frac{\theta}{2} + \sum_{n=1}^{\infty}\frac{nh^{2n}}{1-h^{2n}}\cos n\theta.$$

Therefore,

$$\phi_1'(\pi) = -\frac{1}{8} + \sum_{n=1}^{\infty}\frac{nh^{2n}}{1-h^{2n}}(-1)^n = -\frac{u_1}{8}, \tag{4.13}$$

$$\phi_1'(\pi\tau) = \frac{h}{2(1-h)^2} + \frac{1}{2}\sum_{n=1}^{\infty}\frac{nh^{2n}}{1-h^{2n}}(h^n + h^{-n}) = \sum_{n=1}^{\infty}\frac{nh^n}{1-h^{2n}} = -\frac{u_2}{8} \tag{4.14}$$

and

$$\begin{aligned}\phi_1'(\pi+\pi\tau) &= \frac{-h}{2(1+h)^2} + \frac{1}{2}\sum_{n=1}^{\infty}\frac{nh^{2n}(-1)^n}{1-h^{2n}}(h^n + h^{-n}) \\ &= \sum_{n=1}^{\infty}\frac{(-1)^n nh^n}{1-h^{2n}} = -\frac{u_3}{8},\end{aligned} \tag{4.15}$$

where u_1, u_2 and u_3 are defined by (4.3)–(4.5). If we substitute the results of (4.13)–(4.15) into (4.10)–(4.12), we deduce that the functions u_1, u_2 and u_3 satisfy Halphen's system of differential equations (4.2).

For future reference, we note from (4.4) and (4.5) that replacing h with $-h$ interchanges u_2 and u_3, that is,

$$u_2(-h) = u_3(h). \tag{4.16}$$

4.2 Jacobi's identities and sums of two and four squares

Subtracting one of Halphen's differential equations from another, we obtain

$$h\frac{d}{dh}(u_1 - u_3) = u_2(u_1 - u_3), \tag{4.17}$$

$$h\frac{d}{dh}(u_1 - u_2) = u_3(u_1 - u_2), \tag{4.18}$$

$$h\frac{d}{dh}(u_3 - u_2) = u_1(u_3 - u_2). \tag{4.19}$$

The first of these gives

$$h\frac{d}{dh}\log(u_1 - u_3) = u_2 = -8\sum_{n=1}^{\infty}\frac{nh^n}{1-h^{2n}} = -4\sum_{n=1}^{\infty}\left(\frac{nh^n}{1+h^n} + \frac{nh^n}{1-h^n}\right).$$

Dividing by h and integrating gives

$$u_1 - u_3 = k\prod_{n=1}^{\infty}\frac{(1-h^n)^4}{(1+h^n)^4},$$

where k is a constant of integration whose value can be determined by letting $h = 0$; we find that $k = 1$. The result is

$$u_1 - u_3 = 1 + 8\sum_{n=1}^{\infty}\frac{(-1)^n nh^n}{1+h^n} = \prod_{n=1}^{\infty}\frac{(1-h^n)^4}{(1+h^n)^4}. \tag{4.20}$$

In the same way, equations (4.18) and (4.19), respectively, lead to

$$u_1 - u_2 = 1 + 8\sum_{n=1}^{\infty}\frac{nh^n}{1+(-h)^n} = \prod_{n=1}^{\infty}\frac{(1-(-h)^n)^4}{(1+(-h)^n)^4} \tag{4.21}$$

and

$$u_3 - u_2 = 16\sum_{n=1}^{\infty}\frac{(2n-1)h^{2n-1}}{1-h^{4n-2}} = 16h\prod_{n=1}^{\infty}\frac{(1-h^{4n})^4}{(1-h^{4n-2})^4}. \tag{4.22}$$

Using Euler's identity for infinite products (2.42) in (4.20)–(4.22), we obtain the classic Jacobi products

$$u_1 - u_3 = \prod_{n=1}^{\infty}(1-h^{2n})^4(1-h^{2n-1})^8, \tag{4.23}$$

$$u_1 - u_2 = \prod_{n=1}^{\infty}(1-h^{2n})^4(1+h^{2n-1})^8, \tag{4.24}$$

and

$$u_3 - u_2 = 16h \prod_{n=1}^{\infty} (1-h^{2n})^4 (1+h^{2n})^8. \tag{4.25}$$

Taking the product of (4.23), (4.24) and (4.25) and again using Euler's identity (2.42), we obtain

$$(u_1 - u_3)(u_1 - u_2)(u_3 - u_2) = 16h \prod_{n=1}^{\infty} (1-h^{2n})^{12}. \tag{4.26}$$

From (4.23)–(4.25) we also deduce beautiful identity

$$\prod_{n=1}^{\infty} (1+h^{2n-1})^8 - \prod_{n=1}^{\infty} (1-h^{2n-1})^8 = 16h \prod_{n=1}^{\infty} (1+h^{2n})^8. \tag{4.27}$$

There are two other beautiful identities of Jacobi connected with (4.27)—see (4.36) and (4.37), below—which we will now derive. We put $\theta = \pi/2$ in (1.5) to obtain

$$\left(\frac{1}{4} - \sum_{n=1}^{\infty} \frac{(-1)^n q^{2n-1}}{1-q^{2n-1}}\right)^2 \tag{4.28}$$
$$= \frac{1}{16} + \sum_{n=1}^{\infty} \frac{(-1)^n q^{2n}}{(1-q^{2n})^2} + \frac{1}{2}\sum_{n=1}^{\infty} \frac{nq^n}{1-q^n} - \sum_{n=1}^{\infty} \frac{(-1)^n nq^{2n}}{1-q^{2n}}.$$

If we expand as a double series and interchange the order of summation, we obtain

$$\sum_{n=1}^{\infty} \frac{(-1)^n q^{2n}}{(1-q^{2n})^2} = -\sum_{m=1}^{\infty} \frac{mq^{2m}}{1+q^{2m}} = 2\sum_{m=1}^{\infty} \frac{mq^{4m}}{1-q^{4m}} - \sum_{m=1}^{\infty} \frac{mq^{2m}}{1-q^{2m}}.$$

Therefore, (4.28) simplifies to

$$\left(\frac{1}{4} - \sum_{n=1}^{\infty} \frac{(-1)^n q^{2n-1}}{1-q^{2n-1}}\right)^2 = \frac{1}{16} + \frac{1}{2}\sum_{n=1}^{\infty} \frac{nq^n}{1-q^n} - 2\sum_{n=1}^{\infty} \frac{nq^{4n}}{1-q^{4n}}.$$

Now multiply by 16 and replace q with h, to finally obtain Ramanujan's formula:

$$\left(1 + 4\sum_{n=1}^{\infty}\left(\frac{h^{4n-3}}{1-h^{4n-3}} - \frac{h^{4n-1}}{1-h^{4n-1}}\right)\right)^2 = \frac{1}{3}\left(4P(h^4) - P(h)\right), \tag{4.29}$$

where P is Ramanujan's Eisenstein series defined by (1.34). We also mention that putting $\theta = 2\pi/3$ in (1.5) yields another of Ramanujan's beautiful identities:

$$\left(1 + 6\sum_{n=1}^{\infty}\left(\frac{h^{3n-2}}{1-h^{3n-2}} - \frac{h^{3n-1}}{1-h^{3n-1}}\right)\right)^2 = \frac{1}{2}\left(3P(h^3) - P(h)\right). \tag{4.30}$$

Next, by (4.21) we have

$$
\begin{aligned}
u_1 - u_2 &= 1 + 8\sum_{n=1}^{\infty} \frac{2nh^{2n}}{1+h^{2n}} + 8\sum_{n=1}^{\infty} \frac{(2n-1)h^{2n-1}}{1-h^{2n-1}} \\
&= 1 + 16\sum_{n=1}^{\infty} \left(\frac{nh^{2n}}{1-h^{2n}} - \frac{2nh^{4n}}{1-h^{4n}}\right) + 8\sum_{n=1}^{\infty} \left(\frac{nh^{n}}{1-h^{n}} - \frac{2nh^{2n}}{1-h^{2n}}\right) \\
&= \frac{1}{3}\left(4P(h^4) - P(h)\right).
\end{aligned} \tag{4.31}
$$

By (4.24), and by taking $z = 1$ in Jacobi's triple product identity (2.36), we obtain

$$
u_1 - u_2 = \prod_{n=1}^{\infty} (1-h^{2n})^4 (1+h^{2n-1})^8 = \left(\sum_{n=-\infty}^{\infty} h^{n^2}\right)^4. \tag{4.32}
$$

By (4.29), (4.31) and (4.32), we deduce

$$
(u_1 - u_2)^{1/2} = \left(\sum_{n=-\infty}^{\infty} h^{n^2}\right)^2 = 1 + 4\sum_{n=1}^{\infty} \left(\frac{h^{4n-3}}{1-h^{4n-3}} - \frac{h^{4n-1}}{1-h^{4n-1}}\right) \tag{4.33}
$$

and

$$
u_1 - u_2 = \left(\sum_{n=-\infty}^{\infty} h^{n^2}\right)^4 = \frac{1}{3}\left(4P(h^4) - P(h)\right). \tag{4.34}
$$

The identities (4.33) and (4.34) can be used to give formulas for the number of representations of any natural number n as the sum of two squares and four squares, respectively, in terms of the divisors of n. Equating coefficients of q^n in (4.33), we find that the number of ways of expressing a positive integer n as a sum of two squares equals

$4 \times$ (the number of divisors of n of the form $4j+1$

minus the number of divisors of n of the form $4j+3$).

By comparing coefficients of q^n in (4.34), it follows that the number of representations of a positive integer n by a sum of four squares is

$8 \times$ the sum of the divisors of n which are not multiples of 4.

If we replace h with $-h$ in (4.33) and add the two results, we obtain on using (4.16),

$$
(u_1 - u_2)^{1/2} + (u_1 - u_3)^{1/2} = 2 - 8\sum_{n=1}^{\infty} \frac{(-1)^n h^{4n-2}}{1-h^{4n-2}} = 2\left(u_1(h^2) - u_2(h^2)\right). \tag{4.35}
$$

Now express each side as an infinite product, using (4.23) and (4.24), and simplify, to obtain another identity of Jacobi:

$$\prod_{n=1}^{\infty}(1+h^{2n-1})^4+\prod_{n=1}^{\infty}(1-h^{2n-1})^4=2\prod_{n=1}^{\infty}(1+h^{2n})^2(1+h^{4n-2})^4. \quad (4.36)$$

Dividing (4.27) by (4.36), we obtain Jacobi's other formula

$$\prod_{n=1}^{\infty}(1+h^{2n-1})^4-\prod_{n=1}^{\infty}(1-h^{2n-1})^4=8h\prod_{n=1}^{\infty}(1+h^{2n})^2(1+h^{4n})^4. \quad (4.37)$$

We take this opportunity to insert here the notation used for the transcendentals encountered here by Ramanujan in his notebooks. The functions φ and ψ are defined by

$$\varphi(h)=\sum_{n=-\infty}^{\infty}h^{n^2} \qquad \text{and} \qquad \psi(h)=\sum_{n=0}^{\infty}h^{n(n+1)/2}. \quad (4.38)$$

By Jacobi's triple product identity (2.36) and Gauss' result (2.43), we have the infinite product representations

$$\varphi(h)=\prod_{n=1}^{\infty}(1-h^{2n})(1+h^{2n-1})^2 \qquad \text{and} \qquad \psi(h)=\prod_{n=1}^{\infty}(1-h^n)(1+h^n)^2. \quad (4.39)$$

These may be compared with the infinite products in (4.23)–(4.25) to give

$$(u_1-u_2)^{1/4}=\varphi(h),$$

$$(u_1-u_3)^{1/4}=\varphi(-h),$$

and

$$(u_3-u_2)^{1/4}=2h^{1/4}\psi(h^2).$$

The roots involved here are uniquely defined. Generally the first coefficient of the expansion in powers of h will be real and positive. Jacobi's identity (4.27) is equivalent to

$$\varphi^4(h)-\varphi^4(-h)=16h\psi^4(h^2), \quad (4.40)$$

which may be written in the explicit form

$$\left(\sum_{n=-\infty}^{\infty}q^{n^2}\right)^4=\left(\sum_{n=-\infty}^{\infty}(-1)^n q^{n^2}\right)^4+\left(\sum_{n=-\infty}^{\infty}q^{(n+\frac{1}{2})^2}\right)^4. \quad (4.41)$$

The transcendentals considered so far are

$$u_1=-8\phi_1'(\pi), \qquad u_2=-8\phi_1'(\pi\tau) \quad \text{and} \quad u_3=-8\phi_1'(\pi+\pi\tau). \quad (4.42)$$

By (1.28) and (1.36), the Weierstrass $\wp$ function is given by

$$\wp(\theta) = -2\phi_1'(\theta) - \frac{P}{12} \tag{4.43}$$

where

$$P = P(q) = 1 - 24\sum_{n=1}^{\infty} \frac{nq^n}{1-q^n} = 1 - 24\sum_{n=1}^{\infty} \frac{nh^{2n}}{1-h^{2n}}.$$

We now define the classical Weierstrass invariants e_1, e_2 and e_3 by

$$e_1 = \wp(\pi), \qquad e_2 = \wp(\pi\tau) \quad \text{and} \quad e_3 = \wp(\pi + \pi\tau). \tag{4.44}$$

By (4.42), (4.43) and (4.44), and using the fact that $q = h^2$, we have

$$e_j = \frac{u_j}{4} - \frac{P(h^2)}{12}, \qquad j = 1, 2, 3, \tag{4.45}$$

so the properties of the e_j follow from the corresponding properties of the u_j, and vice versa. From (1.34) and (4.3)–(4.5) we readily deduce the series expansions

$$e_1 = e_1(h) = \frac{1}{6} + 4\sum_{n=1}^{\infty} \frac{(2n-1)h^{4n-2}}{1-h^{4n-2}} = \frac{1}{6} + 4h^2 + O(h^3), \tag{4.46}$$

$$e_2 = e_2(h) = -\frac{1}{12} - 2\sum_{n=1}^{\infty} \frac{(2n-1)h^{2n-1}}{1-h^{2n-1}} = -\frac{1}{12} - 2h - 2h^2 + O(h^3), \tag{4.47}$$

and

$$e_3 = e_3(h) = -\frac{1}{12} + 2\sum_{n=1}^{\infty} \frac{(2n-1)h^{2n-1}}{1+h^{2n-1}} = -\frac{1}{12} + 2h - 2h^2 + O(h^3). \tag{4.48}$$

From (4.46)–(4.48) we note that

$$e_1(h) = -2e_2(h^2), \qquad e_3(h) = e_2(-h)$$

and also the more significant property

$$e_1 + e_2 + e_3 = 0. \tag{4.49}$$

From (4.45) and (4.49), we deduce that

$$u_1(h) + u_2(h) + u_3(h) = P(h^2). \tag{4.50}$$

This can also be deduced from the definitions (4.3)–(4.5) by direct series manipulations.

We can use Ramanujan's identity (1.5) in the form (1.22) to obtain other Lambert series for e_1, e_2, e_3 which are also interesting. Using (1.22), (1.28) and (3.30) we have

$$\wp(\theta) = 2(\phi_2(0) - \phi_1'(\theta)) = 4(\phi_1^2(\theta) - \phi_2(\theta)) = 4(S_1^2(\theta) - C_2(\theta)).$$

Successively substitute $\theta = \pi$, $\theta = \pi\tau$ and $\theta = \pi + \pi\tau$ and use (3.31) and (3.32) to obtain

$$e_1 = \frac{1}{6} + 4\sum_{n=1}^{\infty} \frac{nh^{2n}}{1+h^{2n}} \tag{4.51}$$

$$e_2 = -\frac{1}{12} - 2\sum_{n=1}^{\infty} \frac{nh^n}{1+h^n} \tag{4.52}$$

and

$$e_3 = -\frac{1}{12} - 2\sum_{n=1}^{\infty} \frac{(-1)^n nh^n}{1+(-h)^n}. \tag{4.53}$$

Prof. V. R. Thiruvenkatachar drew my attention to this. He also proved directly the equivalence of (4.46)–(4.48) with (4.51)–(4.53) using the identity

$$\sum_{n=1}^{\infty} \frac{(2n-1)t^{2n-1}}{1-t^{2n-1}} = \sum_{n=1}^{\infty} \frac{nt^n}{1+t^n}$$

which is obtained by logarithmically differentiating the Euler identity (2.42).

4.3 Quadratic transformations

Quadratic transformation of the elliptic transcendentals follows easily from the formulas derived above using the Halphen differential equations. Consider the transformation $h \to h^2$ on the functions u_1, u_2, u_3, and let

$$U_j(h) = u_j(h^2), \qquad j = 1, 2, 3. \tag{4.54}$$

From (4.35) we have

$$(u_1 - u_2)^{1/2} + (u_1 - u_3)^{1/2} = 2\,(U_1 - U_2)^{1/2}. \tag{4.55}$$

Now

$$(u_1 - u_2)^{1/2} - (u_1 - u_3)^{1/2} = \frac{(u_1 - u_2) - (u_1 - u_3)}{(u_1 - u_2)^{1/2} + (u_1 - u_3)^{1/2}} = \frac{u_3 - u_2}{2(U_1 - U_2)^{1/2}}.$$

Comparing the numerator and denominator on the right hand side with the infinite products in (4.21) and (4.22), we get

$$\begin{aligned}(u_1-u_2)^{1/2}-(u_1-u_3)^{1/2} &= 8h\prod_{n=1}^{\infty}\frac{(1-h^{2n})^4(1+h^{2n})^8}{(1-h^{4n})^2(1+h^{4n-2})^4} \qquad (4.56)\\ &= 8h\prod_{n=1}^{\infty}(1-h^{4n})^2(1+h^{4n})^4\\ &= 2(U_3-U_2)^{1/2}.\end{aligned}$$

From (4.4) and (4.5) we get

$$u_2+u_3=-32\sum_{n=1}^{\infty}\frac{nh^{2n}}{1-h^{4n}}=4U_2. \tag{4.57}$$

Hence, from (4.55), (4.56) and (4.57) we deduce the quadratic transformation formulas for the Halphen functions:

$$U_1=\frac{1}{2}\left(u_1+\sqrt{(u_1-u_2)(u_1-u_3)}\right), \tag{4.58}$$

$$U_2=\frac{1}{4}(u_2+u_3), \tag{4.59}$$

and

$$U_3=\frac{1}{2}\left(u_1-\sqrt{(u_1-u_2)(u_1-u_3)}\right). \tag{4.60}$$

As before, the signs of the square roots are determined uniquely by requiring that the first coefficient of the expansion in powers of h will be real and positive.

Solving for u_1, u_2, u_3 in terms of U_1, U_2, U_3 we obtain

$$u_1=U_1+U_3, \tag{4.61}$$

$$u_2=2\left(U_2-\sqrt{(U_1-U_2)(U_3-U_2)}\right), \tag{4.62}$$

and

$$u_3=2\left(U_2+\sqrt{(U_1-U_2)(U_3-U_2)}\right). \tag{4.63}$$

From these we can obtain the quadratic transformations of e_1, e_2, e_3 and P.

We now define the Legendre modular function λ by

$$\lambda=\lambda(h)=\frac{e_3-e_2}{e_1-e_2}=\frac{u_3-u_2}{u_1-u_2}. \tag{4.64}$$

By (4.23)–(4.25) we have the representations by infinite products given by

$$\lambda(h) = 16h \prod_{n=1}^{\infty} \frac{(1+h^{2n})^8}{(1+h^{2n-1})^8}, \tag{4.65}$$

and

$$1-\lambda(h) = \prod_{n=1}^{\infty} \frac{(1-h^{2n-1})^8}{(1+h^{2n-1})^8}. \tag{4.66}$$

One derives the quadratic transformation of λ from (4.61)–(4.63). It is easily seen to be the following:

$$\lambda = \frac{4\sqrt{\Lambda}}{(1+\sqrt{\Lambda})^2} \tag{4.67}$$

where $\lambda = \lambda(h)$ and $\Lambda = \lambda(h^2)$. Here, $\sqrt{\Lambda}$ is uniquely defined as the series beginning with $\Lambda = 4h + O(h^2)$.

If the first n coefficients in the expansion of $\lambda(h)$ are known, then the next n coefficients can be determined using a procedure described by Ramanujan in his notebooks [92, Ch. 17, Entry 2], which we will now describe. Suppose

$$\lambda = a_1 h + a_2 h^2 + \cdots + a_n h^n + O(h^{n+1}), \qquad a_1 \neq 0,$$

where $a_1, a_2, \ldots, a_n$ are known. Such an expansion clearly exists, and in fact by (4.65) we have $a_1 = 16$. Then

$$\Lambda = a_1 h^2 + a_2 h^4 + \cdots + a_n h^{2n} + O(h^{2n+2})$$

and so

$$\sqrt{\Lambda} = \sqrt{a_1}\, h \left(1 + \frac{a_2}{a_1}h^2 + \frac{a_3}{a_1}h^4 + \cdots + \frac{a_n}{a_1}h^{2n-2} + O(h^{2n})\right)^{1/2}$$

$$= b_1 h + b_2 h^3 + b_3 h^5 + \cdots + b_n h^{2n-1} + O(h^{2n+1}),$$

where $b_1, b_2, \ldots, b_n$ can be computed from the numbers $a_1, a_2, \ldots, a_n$ using the binomial expansion for the square root. Thus,

$$\lambda = \frac{4\sqrt{\Lambda}}{(1+\sqrt{\Lambda})^2} = \frac{4(b_1 h + b_2 h^3 + b_3 h^5 + \cdots + b_n h^{2n-1} + O(h^{2n+1}))}{(1 + b_1 h + b_2 h^3 + b_3 h^5 + \cdots + b_n h^{2n-1} + O(h^{2n+1}))^2}.$$

It follows that if the first n coefficients of the expansion of λ are known, then these can be used to determine the first $2n$ coefficients. This amounts to the interesting fact that the function $\lambda(h)$, defined as a power series with its constant term 0 and not identically zero, is determined uniquely by (4.67).

4.4 The hypergeometric function ${}_2F_1(1/2, 1/2; 1; x)$: second method

We now derive the connection with the hypergeometric function by another method.

We will need the classical factorization given by

$$4\wp^3(\theta) - \frac{Q}{12}\wp(\theta) - \frac{R}{216} = 4\left(\wp(\theta) - e_1\right)\left(\wp(\theta) - e_2\right)\left(\wp(\theta) - e_3\right), \tag{4.68}$$

where e_1, e_2 and e_3 are given by (4.46)–(4.48), and

$$Q = 1 + 240\sum_{n=1}^{\infty}\frac{n^3h^{2n}}{1-h^{2n}} \qquad \text{and} \qquad R = 1 - 504\sum_{n=1}^{\infty}\frac{n^5h^{2n}}{1-h^{2n}}. \tag{4.69}$$

This may be proved in the standard way, as follows. First, by (1.18)–(1.20) and (1.28), it follows that $\wp'(\theta)$ is an odd elliptic function with periods 2π and $2\pi\tau$. Next, since $\phi_1(\theta)$ is analytic at the half-periods $\theta = \pi$, $\pi\tau$ and $\pi + \pi\tau$, we have that $\wp'(\theta)$ is also analytic at these points. It follows that $\wp'(\theta)$ is zero at each of the half-periods, and therefore by (1.42), the Weierstrassian cubic

$$4\wp^3(\theta) - \frac{Q}{12}\wp(\theta) - \frac{R}{216}$$

is zero when $\theta = \pi$, $\pi\tau$ or $\pi + \pi\tau$. We could deduce the factorization

$$4\wp^3(\theta) - \frac{Q}{12}\wp(\theta) - \frac{R}{216} = 4\left(\wp(\theta) - \wp(\pi)\right)\left(\wp(\theta) - \wp(\pi\tau)\right)\left(\wp(\theta) - \wp(\pi + \pi\tau)\right)$$

if we knew that $\wp(\pi)$, $\wp(\pi\tau)$ and $\wp(\pi + \pi\tau)$ were all distinct; but these are just e_1, e_2 and e_3, respectively, and they are distinct by (4.26) and (4.45). Thus, the factorization of the Weierstrass cubic given by (4.68) has been proved.

From the factorization formula (4.68) we deduce that

$$e_1 + e_2 + e_3 = 0, \tag{4.70}$$

$$e_1e_2 + e_2e_3 + e_3e_1 = -\frac{Q}{48}, \tag{4.71}$$

and

$$e_1e_2e_3 = \frac{R}{864}, \tag{4.72}$$

where e_1, e_2 and e_3 are given by (4.46)–(4.48), and Q and R are given by (4.69). We observe that

$$\begin{aligned}
1-\lambda+\lambda^2 &= 1-\frac{e_3-e_2}{e_1-e_2}+\frac{(e_3-e_2)^2}{(e_1-e_2)^2}\\
&= \frac{1}{(e_1-e_2)^2}\left\{e_1^2+e_2^2+e_3^2-e_1e_2-e_2e_3-e_3e_1\right\}\\
&= \frac{1}{(e_1-e_2)^2}\left\{(e_1+e_2+e_3)^2-3(e_1e_2+e_2e_3+e_3e_1)\right\}\\
&= \frac{Q}{16(e_1-e_2)^2}
\end{aligned}$$

and

$$\begin{aligned}
&(1+\lambda)(1-2\lambda)(2-\lambda)\\
&= \frac{1}{(e_1-e_2)^3}(e_1-2e_2+e_3)(e_1+e_2-2e_3)(2e_1-e_2-e_3)\\
&= \frac{27e_1e_2e_3}{(e_1-e_2)^3}\\
&= \frac{R}{32(e_1-e_2)^3}.
\end{aligned}$$

Let $\mu = 2\sqrt{e_1-e_2}$. Then, we have proved

$$Q=\mu^4(1-\lambda+\lambda^2) \qquad \text{and} \qquad R=\mu^6(1+\lambda)(1-2\lambda)(1-\lambda/2), \tag{4.73}$$

where

$$Q = 1+240\sum_{n=1}^{\infty}\frac{n^3h^{2n}}{1-h^{2n}}, \qquad R=1-504\sum_{n=1}^{\infty}\frac{n^5h^{2n}}{1-h^{2n}}, \tag{4.74}$$

$$\mu = 2\sqrt{e_1-e_2} = \prod_{n=1}^{\infty}(1-h^{2n})^2(1+h^{2n-1})^4 = \left(\sum_{n=-\infty}^{\infty} h^{n^2}\right)^2 \tag{4.75}$$

and

$$\lambda = \frac{e_3-e_2}{e_1-e_2} = 16h\prod_{n=1}^{\infty}\frac{(1+h^{2n})^8}{(1+h^{2n-1})^8}. \tag{4.76}$$

We reiterate that $h=\exp(i\pi\tau)$ and $q=\exp(2\pi i\tau)=h^2$.

It follows from (4.73) that λ satisfies the sextic equation

$$\frac{Q^3}{Q^3-R^2} = \frac{4}{27}\frac{(1-\lambda+\lambda^2)^3}{\lambda^2(1-\lambda)^2}. \tag{4.77}$$

This has six roots corresponding to the quotients of the mutual differences of the e_j. In fact, the rational function

$$\frac{4}{27}\frac{(1-\lambda+\lambda^2)^3}{\lambda^2(1-\lambda)^2}$$

is invariant under the transformations $\lambda \to 1-\lambda$ and $\lambda \to 1/\lambda$. These transformations generate a group of order 6, namely

$$\lambda \to \lambda,\ 1-\lambda,\ \frac{1}{\lambda},\ \frac{1}{1-\lambda},\ 1-\frac{1}{\lambda},\ \frac{\lambda}{\lambda-1}.$$

Thus, the six roots of (4.77) are given by

$$\lambda \in \left\{\frac{e_3-e_2}{e_1-e_2},\quad \frac{e_3-e_1}{e_2-e_1},\quad \frac{e_1-e_2}{e_3-e_2},\quad \frac{e_2-e_1}{e_3-e_1},\quad \frac{e_1-e_3}{e_2-e_3},\quad \frac{e_2-e_3}{e_1-e_3}\right\},$$

that is to say

$$\lambda = \frac{e_a-e_b}{e_c-e_b}$$

where (a,b,c) is any permutation of $(1,2,3)$.

We now prove that λ satisfies a third order differential equation with respect to h. Let us use a prime to denote differentiation with respect to $\log h$. That is,

$$f' = \frac{df}{d\log h} = h\frac{df}{dh}.$$

We logarithmically differentiate the Legendre modular function defined by (4.64) and apply the Halphen differential equations to get

$$\frac{\lambda'}{\lambda} = \frac{u_3'-u_2'}{u_3-u_2} - \frac{u_1'-u_2'}{u_1-u_2} = u_1-u_3. \tag{4.78}$$

Similarly, logarithmically differentiating $1-\lambda$ gives

$$\frac{\lambda'}{1-\lambda} = u_3-u_2. \tag{4.79}$$

Logarithmically differentiating these again gives

$$\frac{\lambda''}{\lambda'} - \frac{\lambda'}{\lambda} = u_2 \tag{4.80}$$

and

$$\frac{\lambda''}{\lambda'} + \frac{\lambda'}{1-\lambda} = u_1, \tag{4.81}$$

and adding (4.79) and (4.80) gives

$$\frac{\lambda''}{\lambda'} - \frac{\lambda'}{\lambda} + \frac{\lambda'}{1-\lambda} = u_3. \tag{4.82}$$

Differentiating (4.80) and (4.81) and adding we obtain

$$\left(\frac{2\lambda''}{\lambda'} - \frac{\lambda'}{\lambda} + \frac{\lambda'}{1-\lambda}\right)' = (u_1 + u_2)' = u_1 u_2, \tag{4.83}$$

while multiplying (4.80) and (4.81) gives

$$\left(\frac{\lambda''}{\lambda'} + \frac{\lambda'}{1-\lambda}\right)\left(\frac{\lambda''}{\lambda'} - \frac{\lambda'}{\lambda}\right) = u_1 u_2. \tag{4.84}$$

Equating (4.83) and (4.84) and simplifying we obtain

$$2\lambda''' - \frac{3\lambda''^2}{\lambda'} + \frac{\lambda'^3(1-\lambda+\lambda^2)}{\lambda^2(1-\lambda)^2} = 0. \tag{4.85}$$

This is the third order differential equation satisfied by λ.

By (4.66), $1-\lambda$ is a decreasing function as h increases from 0 to 1. Therefore, the inverse function $h = h(\lambda)$ exists. Let t be defined by

$$t = t(\lambda) = \log h(\lambda).$$

It is interesting to obtain the differential equation of the function t with respect to λ. Let

$$\dot{t} = \frac{dt}{d\lambda}, \qquad \ddot{t} = \frac{d^2t}{d\lambda^2}, \quad \text{and} \quad \dddot{t} = \frac{d^3t}{d\lambda^3}.$$

Then

$$\dot{t} = \frac{dt}{d\lambda} = \frac{d\log h}{d\lambda} = \frac{1}{h}\frac{dh}{d\lambda} = \frac{1}{\lambda'}, \qquad \text{and} \qquad \lambda' = \frac{1}{\dot{t}}.$$

Differentiating gives

$$\lambda'' = -\frac{\ddot{t}}{\dot{t}^3} \qquad \text{and} \qquad \lambda''' = \frac{1}{\dot{t}}\left(-\frac{\dddot{t}}{\dot{t}^3} + \frac{3\ddot{t}^2}{\dot{t}^4}\right).$$

Substituting into (4.85) and simplifying gives

$$\frac{2\dddot{t}}{\dot{t}} - \frac{3\ddot{t}^2}{\dot{t}^2} = \frac{\lambda^2-\lambda+1}{\lambda^2(1-\lambda)^2}. \tag{4.86}$$

The Schwarzian derivative of t with respect to λ is defined by

$$\{t,\lambda\} = \frac{\dddot{t}}{\dot{t}} - \frac{3}{2}\left(\frac{\ddot{t}}{\dot{t}}\right)^2, \tag{4.87}$$

so in this notation, (4.86) becomes

$$2\{t,\lambda\} = \frac{\lambda^2-\lambda+1}{\lambda^2(1-\lambda)^2}. \tag{4.88}$$

Since

$$\left\{\frac{at+b}{ct+d}, \lambda\right\} = \{t, \lambda\}$$

we obtain that (4.85) has as general solution $\lambda\left(\frac{at+b}{ct+d}\right)$, $ad - bc \neq 0$, a, b, c, d being constants.

We now use the Halphen differential equations to derive the differential equation satisfied by μ with respect to λ. As before, let $t = \log h = i\pi\tau$. From (4.80) and (4.81),

$$\mu^2 = 4(e_1 - e_2) = u_1 - u_2 = \frac{\lambda'}{1-\lambda} + \frac{\lambda'}{\lambda} = \frac{\lambda'}{\lambda(1-\lambda)}.$$

Therefore

$$\mu^2\lambda(1-\lambda) = \lambda' = \frac{d\lambda}{dt} = \left(\frac{dt}{d\lambda}\right)^{-1},$$

and so

$$\mu\lambda^{1/2}(1-\lambda)^{1/2} = \left(\frac{dt}{d\lambda}\right)^{-1/2}. \tag{4.89}$$

Differentiating with respect to λ gives

$$\frac{d}{d\lambda}\left(\mu\lambda^{1/2}(1-\lambda)^{1/2}\right) = -\frac{1}{2}\left(\frac{dt}{d\lambda}\right)^{-3/2}\frac{d^2t}{d\lambda^2}.$$

Differentiating again and using (4.88) and (4.89) gives

$$\begin{aligned}\frac{d^2}{d\lambda^2}\left(\mu\lambda^{1/2}(1-\lambda)^{1/2}\right) &= \frac{3}{4}\left(\frac{dt}{d\lambda}\right)^{-5/2}\left(\frac{d^2t}{d\lambda^2}\right)^2 - \frac{1}{2}\left(\frac{dt}{d\lambda}\right)^{-3/2}\frac{d^3t}{d\lambda^3}\\ &= -\frac{1}{2}\left(\frac{dt}{d\lambda}\right)^{-1/2}\{t, \lambda\}\\ &= -\frac{\mu}{4}\frac{(\lambda^2-\lambda+1)}{\lambda^{3/2}(1-\lambda)^{3/2}}.\end{aligned}$$

Expanding the derivative on the left hand side and simplifying leads to

$$\frac{d}{d\lambda}\left(\lambda(1-\lambda)\frac{d\mu}{d\lambda}\right) = \frac{\mu}{4}. \tag{4.90}$$

This is the hypergeometric differential equation satisfied by μ with respect to λ.

We now obtain the imaginary transformations of u_1, u_2, u_3, and hence also of e_1, e_2 and e_3. Let h and $\underline{h}$ be defined by

$$h = \exp(i\pi\tau) \qquad \text{and} \qquad \underline{h} = \exp(-i\pi/\tau).$$

With this notation, the transformation formulas (2.71)–(2.74) take the form

$$h^{1/12}\prod_{n=1}^{\infty}(1+h^{2n}) = \frac{1}{\sqrt{2}\,\underline{h}^{1/24}}\prod_{n=1}^{\infty}(1-\underline{h}^{2n-1}), \tag{4.91}$$

$$\frac{1}{h^{1/24}}\prod_{n=1}^{\infty}(1-h^{2n-1}) = \sqrt{2}\,\underline{h}^{1/12}\prod_{n=1}^{\infty}(1+\underline{h}^{2n}), \tag{4.92}$$

$$\frac{1}{h^{1/24}}\prod_{n=1}^{\infty}(1+h^{2n-1}) = \frac{1}{\sqrt{2}\,\underline{h}^{1/24}}\prod_{n=1}^{\infty}(1+\underline{h}^{2n-1}) \tag{4.93}$$

and

$$h^{1/12}\prod_{n=1}^{\infty}(1-h^{2n}) = \sqrt{\frac{i}{\tau}}\,\underline{h}^{1/12}\prod_{n=1}^{\infty}(1-\underline{h}^{2n}). \tag{4.94}$$

Using (4.23)–(4.25) and (4.91)–(4.94) we can immediately write down the imaginary transformations of $u_1 - u_2$, $u_1 - u_3$ and $u_3 - u_2$. For $r = 1$, 2 or 3, write u_r and $\underline{u}_r$ for $u_r(h)$ and $u_r(\underline{h})$, respectively. Then we have:

$$u_1 - u_2 = -\frac{1}{\tau^2}(\underline{u}_1 - \underline{u}_2), \tag{4.95}$$

$$u_1 - u_3 = -\frac{1}{\tau^2}(\underline{u}_3 - \underline{u}_2), \tag{4.96}$$

and

$$u_3 - u_2 = -\frac{1}{\tau^2}(\underline{u}_1 - \underline{u}_3). \tag{4.97}$$

These also imply

$$(u_1-u_3)^2(u_1-u_2)^2(u_3-u_2)^2 = \frac{1}{\tau^{12}}(\underline{u}_1-\underline{u}_3)^2(\underline{u}_1-\underline{u}_2)^2(\underline{u}_3-\underline{u}_2)^2. \tag{4.98}$$

This is equivalent to the transformation of the discriminant

$$Q^3(h^2) - R^2(h^2) = \frac{1}{\tau^{12}}(Q^3(\underline{h}^2) - R^2(\underline{h}^2))$$

which may be derived in another way using (2.57).

We now obtain the imaginary transformations of u_1, u_2 and u_3. Now

$$h\frac{d}{dh} = \frac{1}{\pi i}\frac{d}{d\tau}, \qquad \text{and} \qquad \underline{h}\frac{d}{d\underline{h}} = \frac{\tau^2}{\pi i}\frac{d}{d\tau},$$

and so

$$h\frac{d}{dh} = \frac{1}{\tau^2}\underline{h}\frac{d}{d\underline{h}}.$$

Using these together with (4.17)–(4.19) and (4.95)–(4.97), we have

$$\begin{aligned} u_1 &= h\frac{d}{dh}\log(u_3-u_2) \\ &= h\frac{d}{dh}\log\left(-\frac{1}{\tau^2}\left(\underline{u}_1-\underline{u}_3\right)\right) \\ &= \frac{1}{\pi i}\frac{d}{d\tau}\log\left(\frac{1}{\tau^2}\right)+\frac{1}{\tau^2}\,\underline{h}\,\frac{d}{d\underline{h}}\log(\underline{u}_1-\underline{u}_3) \\ &= \frac{2i}{\pi\tau}+\frac{1}{\tau^2}\,\underline{u}_2, \end{aligned}$$

and similarly we obtain

$$u_2=\frac{2i}{\pi\tau}+\frac{1}{\tau^2}\,\underline{u}_1 \qquad \text{and} \qquad u_3=\frac{2i}{\pi\tau}+\frac{1}{\tau^2}\,\underline{u}_3.$$

Thus,

$$u_1\left(e^{-i\pi/\tau}\right)=\tau^2u_2(e^{i\pi\tau})+\frac{2\tau}{\pi i}, \tag{4.99}$$

$$u_2\left(e^{-i\pi/\tau}\right)=\tau^2u_1(e^{i\pi\tau})+\frac{2\tau}{\pi i}, \tag{4.100}$$

and

$$u_3\left(e^{-i\pi/\tau}\right)=\tau^2u_3(e^{i\pi\tau})+\frac{2\tau}{\pi i}. \tag{4.101}$$

Adding (4.99)–(4.101) and applying (4.50), we obtain

$$P\left(e^{-2\pi i/\tau}\right)=\tau^2P(e^{2\pi i\tau})+\frac{6\tau}{\pi i} \tag{4.102}$$

which we derived earlier in (2.58) by another method.

The imaginary transformations of e_1, e_2 and e_3 follow immediately from (4.45) and (4.99)–(4.102). We find that

$$e_1\left(e^{-i\pi/\tau}\right)=e_2(e^{i\pi\tau}), \qquad e_2\left(e^{-i\pi/\tau}\right)=e_1(e^{i\pi\tau})$$

and

$$e_3\left(e^{-i\pi/\tau}\right)=e_3(e^{i\pi\tau}).$$

The imaginary transformation of λ is easily obtained from the results above. We have

$$\lambda(h)=\frac{u_3-u_2}{u_1-u_2}=\frac{\underline{u}_1-\underline{u}_3}{\underline{u}_1-\underline{u}_2}=1-\lambda\left(\underline{h}\right). \tag{4.103}$$

We now consider briefly the transformation $\tau\to\tau+1$. This changes h to $-h$. From (4.3)–(4.5) we find that

$$u_1(-h)=u_1(h), \tag{4.104}$$

$$u_2(-h) = u_3(h), \tag{4.105}$$

and

$$u_3(-h) = u_2(h) \tag{4.106}$$

Thus

$$\lambda(h) = \frac{u_3 - u_2}{u_1 - u_2} = \frac{u_2(-h) - u_3(-h)}{u_1(-h) - u_3(-h)} = 1 - \frac{1}{1 - \lambda(-h)}. \tag{4.107}$$

The transformations $\tau \to \tau + 1$ and $\tau \to -1/\tau$ enable us to study the transformation $\tau \to \dfrac{a\tau + b}{c\tau + d}$, where a, b, c, d are integers satisfying $ad - bc = 1$.

The solution of (4.90) is given by

$$\mu = A\,{}_2F_1(1/2, 1/2; 1; \lambda) + B\,{}_2F_1(1/2, 1/2; 1; 1 - \lambda) \tag{4.108}$$

for some constants A and B. From (4.75) and (4.76) we have the expansions

$$\mu \ = \ 2\sqrt{e_1 - e_2} \ = \ \prod_{n=1}^{\infty}(1 - h^{2n})^2(1 + h^{2n-1})^4 \ = \ 1 + 4h + O(h^2)$$

and

$$\lambda \ = \ \frac{e_3 - e_2}{e_1 - e_2} \ = \ 16h\prod_{n=1}^{\infty}\frac{(1 + h^{2n})^8}{(1 + h^{2n-1})^8} \ = \ 16h + O(h^2).$$

Thus, as $h \to 0^+$, $\mu \to 1$ and $\lambda \to 0^+$. The values of the constants A and B may be determined by letting $h \to 0^+$ in (4.108). By (3.68),

$$\lim_{\lambda \to 0^+} {}_2F_1(1/2, 1/2; 1; 1 - \lambda) = +\infty.$$

It follows that $A = 1$ and $B = 0$, and so we have proved

$$\mu = {}_2F_1(1/2, 1/2; 1; \lambda). \tag{4.109}$$

In terms of the Halphen functions u_1, u_2 and u_3, this is

$$\sqrt{u_1 - u_2} = {}_2F_1\left(\frac{1}{2}, \frac{1}{2}; 1; \frac{u_3 - u_2}{u_1 - u_2}\right).$$

We replace τ with $-1/\tau$ to get

$$\sqrt{\underline{u}_1 - \underline{u}_2} = {}_2F_1\left(\frac{1}{2}, \frac{1}{2}; 1; \frac{\underline{u}_3 - \underline{u}_2}{\underline{u}_1 - \underline{u}_2}\right).$$

Now apply the transformation formulas (4.99)–(4.101) to get

$$\tau\sqrt{u_2 - u_1} = {}_2F_1\left(\frac{1}{2}, \frac{1}{2}; 1; \frac{u_3 - u_1}{u_2 - u_1}\right),$$

and hence

$$\pm i\tau\mu = {}_2F_1(1/2, 1/2; 1; 1-\lambda). \tag{4.110}$$

The sign may be determined by noting from (4.103) that when $\tau = i$, $\lambda = 1-\lambda$, and comparing (4.109) with (4.110). It follows that

$$-i\tau\mu = {}_2F_1(1/2, 1/2; 1; 1-\lambda). \tag{4.111}$$

Comparing (4.109) with (4.111), we deduce that if λ is given by (4.64) or (4.65), i.e.,

$$\lambda = \frac{u_3(h) - u_2(h)}{u_1(h) - u_2(h)} = 16h \prod_{j=1}^{\infty} \frac{(1+h^{2n})^8}{(1+h^{2n-1})^8}$$

where $h = e^{i\pi\tau}$, then the inverse function is given by

$$\tau = i\, \frac{{}_2F_1(1/2, 1/2; 1; 1-\lambda)}{{}_2F_1(1/2, 1/2; 1; \lambda)},$$

or

$$h = \exp\left(-\pi\, \frac{{}_2F_1(1/2, 1/2; 1; 1-\lambda)}{{}_2F_1(1/2, 1/2; 1; \lambda)}\right).$$

The treatment of the topics in this chapter is independent of the methods used in the preceding chapter. All the results obtained there using the Ramanujan differential equations are obtained here using the Halphen differential equations. In the next chapter we expound another method which is independent of both of these. This is entirely different from the classical treatment. We will also establish the relations

$$x = \lambda(h^2) \qquad \text{and} \qquad z = \mu(h^2)$$

that connect the transcendentals x and z used in Chapter 3 with the λ and μ used in this chapter.

4.5 Notes

The Puiseux-Halphen inequality occurs in Halphen's work in [61, p. 128, (96)]. An interesting and accessible account of the Puiseux-Halphen inequality has been given by A. Weinstein [110].

The system (4.2) and other related systems of nonlinear differential equations have been studied by W. Zudilin [114].

The identities (4.29) and (4.30) appear as the equations numbered (20) and (19), respectively, in Ramanujan's paper [91].

The sum of two squares and sum of four squares identities, (4.33) and (4.34), are due to Jacobi [72]. There are many proofs of these results, and the reader is referred to [16, Sec. 3.8] for some references. For a different approach, see the recent book by K. S. Williams [112].

A good reference for the Schwarzian derivative (4.87) is the book by L. R. Ford [55, pp. 98–101].

Chapter 5

The Weierstrassian Invariants, II

5.1 Parameterizations of Eisenstein series

In Sec. 3.4, we defined x and z by

$$Q(q) = z^4(1+14x+x^2) \qquad \text{and} \qquad R(q) = z^6(1+x)(1-34x+x^2), \quad (5.1)$$

with $(x, z) = (0, 1)$ when $q = 0$. In Sec. 4.4 we defined λ and μ in terms of the Weierstrass invariants e_1, e_2 and e_3 and proved that

$$Q(h^2) = Q(q) = \mu^4(1 - \lambda + \lambda^2)$$

and

$$R(h^2) = R(q) = \mu^6(1 + \lambda)(1 - 2\lambda)(1 - \lambda/2).$$

We will now investigate the relation of x and z to λ and μ. Let us define $\widetilde{Q}$ and $\widetilde{R}$ by

$$\widetilde{Q} = \widetilde{Q}(q) = z^4(1 - x + x^2) \tag{5.2}$$

and

$$\widetilde{R} = \widetilde{R}(q) = z^6(1 + x)(1 - 2x)(1 - x/2), \tag{5.3}$$

where x and z are given above by (5.1). We use a prime to denote differentiation with respect to $\log q$. Then,

$$\frac{\widetilde{Q'}}{\widetilde{Q}} = \frac{4z'}{z} + \frac{(2x-1)x'}{1-x+x^2} \tag{5.4}$$

and

$$\frac{\widetilde{R}'}{\widetilde{R}} = \frac{6z'}{z} + \frac{(3x^2-3x-3/2)x'}{(1+x)(1-2x)(1-x/2)}. \tag{5.5}$$

We note that

$$\widetilde{Q}^3 - \widetilde{R}^2 = \frac{27}{4} z^{12} x^2 (1-x)^2.$$

We now define $\widetilde{P}$ by either

$$\widetilde{P} = \frac{\widetilde{R}}{\widetilde{Q}} + \frac{3\widetilde{Q}'}{2\widetilde{Q}} \tag{5.6}$$

or

$$\widetilde{P} = \frac{\widetilde{Q}^2}{\widetilde{R}} + \frac{\widetilde{R}'}{\widetilde{R}}. \tag{5.7}$$

The two definitions are equivalent because, by (5.2)–(5.5) we have

$$\frac{\widetilde{R}}{\widetilde{Q}} + \frac{3\widetilde{Q}'}{2\widetilde{Q}} = \frac{\widetilde{Q}^2}{\widetilde{R}} + \frac{\widetilde{R}'}{\widetilde{R}} = \frac{6z'}{z} + z^2(1-2x),$$

and so

$$\widetilde{P} = \frac{6z'}{z} + z^2(1-2x). \tag{5.8}$$

We now prove that $\widetilde{P}$, $\widetilde{Q}$ and $\widetilde{R}$ are equal to $P(q^2)$, $Q(q^2)$ and $R(q^2)$, respectively. From (3.60) we have

$$\frac{z'}{z} = \frac{1}{12}\left(P - (1-5x)z^2\right). \tag{5.9}$$

If we combine (5.8) with (5.9), we find that

$$2\widetilde{P} = P + z^2(1+x). \tag{5.10}$$

Differentiating with respect to $\log q$ gives

$$2q\frac{d\widetilde{P}}{dq} = q\frac{dP}{dq} + 2z\,q\frac{dz}{dq}(1+x) + z^2\,q\frac{dx}{dq}.$$

The derivatives on the right hand side may be computed using (2.13), (3.57), (3.63) and (3.64), and the result may be expressed in the form

$$2q\frac{d\widetilde{P}}{dq} = \frac{1}{12}\left(P + z^2(1+x)\right)^2 - \frac{z^4}{3}(1-x+x^2).$$

By (5.2) and (5.10), this is equivalent to

$$q\frac{d\widetilde{P}}{dq} = \frac{1}{6}\left(\widetilde{P}^2 - \widetilde{Q}\right).$$

From (5.6) and (5.7), we have

$$q\frac{d\widetilde{Q}}{dq} = \frac{2}{3}\left(\widetilde{P}\widetilde{Q} - \widetilde{R}\right)$$

and

$$q\frac{d\widetilde{R}}{dq} = \widetilde{P}\widetilde{R} - \widetilde{Q}^2.$$

Therefore, $\widetilde{P}$, $\widetilde{Q}$ and $\widetilde{R}$ satisfy the system of differential equations:

$$q^2\frac{d\widetilde{P}}{dq^2} = \frac{\widetilde{P}^2 - \widetilde{Q}}{12}, \qquad q^2\frac{d\widetilde{Q}}{dq^2} = \frac{\widetilde{P}\widetilde{Q} - \widetilde{R}}{3}, \qquad q^2\frac{d\widetilde{R}}{dq^2} = \frac{\widetilde{P}\widetilde{R} - \widetilde{Q}^2}{2}. \quad (5.11)$$

That is, the functions $\widetilde{P}$, $\widetilde{Q}$ and $\widetilde{R}$ satisfy Ramanujan's differential equations with respect to q^2. Using the definitions (5.2), (5.3) and (5.4), together with the expansions for x and z given by (3.59), we find that

$$\begin{cases} \widetilde{P} = 1 - 24q^2 + O(q^3), \\ \widetilde{Q} = 1 + 240q^2 + O(q^3), \\ \widetilde{R} = 1 - 504q^2 + O(q^3). \end{cases} \quad (5.12)$$

The differential equations in (5.11) together with the initial coefficients in (5.12) will enable one to successively calculate the other coefficients. It follows that

$$\widetilde{P}(q) = P(q^2), \qquad \widetilde{Q}(q) = Q(q^2), \qquad \text{and} \qquad \widetilde{R}(q) = R(q^2). \quad (5.13)$$

We compare the formulas for Q and R in terms λ and μ given by (4.73)–(4.76), with the formulas for $\widetilde{Q}$ and $\widetilde{R}$ in terms of x and z given by (3.57), (3.58) and (5.13), and obtain

$$x = x(q) = \lambda(h^2) \qquad \text{and} \qquad z = z(q) = \mu(h^2). \quad (5.14)$$

We collect below the formulas obtained so far.

Recall that $q = e^{2\pi i\tau}$, where $\mathrm{Im}(\tau) > 0$. Let x and z be defined by

$$Q(q) = z^4(1+14x+x^2) \qquad \text{and} \qquad R(q) = z^6(1+x)(1-34x+x^2), \quad (5.15)$$

where $(x, z) = (0, 1)$ when $q = 0$. Then

$$Q(q^2) = z^4(1 - x + x^2) \qquad \text{and} \qquad R(q^2) = z^6(1 + x)(1 - 2x)(1 - x/2) \quad (5.16)$$

and

$$q\frac{dx}{dq} = z^2x(1 - x). \quad (5.17)$$

In terms of x, we have

$$z = {}_2F_1(1/2, 1/2; 1; x), \quad (5.18)$$

$$\tau = \frac{i}{2}\,\frac{{}_2F_1(1/2,1/2;1;1-x)}{{}_2F_1(1/2,1/2;1;x)}, \tag{5.19}$$

and

$$q = \exp(2\pi i\tau) = \exp\left(-\pi\,\frac{{}_2F_1(1/2,1/2;1;1-x)}{{}_2F_1(1/2,1/2;1;x)}\right).$$

In terms of q, we have

$$x = \frac{u_3(q)-u_2(q)}{u_1(q)-u_2(q)} = 16q\prod_{n=1}^{\infty}\frac{(1+q^{2n})^8}{(1+q^{2n-1})^8}, \tag{5.20}$$

$$1-x = \frac{u_1(q)-u_3(q)}{u_1(q)-u_2(q)} = \prod_{n=1}^{\infty}\frac{(1-q^{2n-1})^8}{(1+q^{2n-1})^8} \tag{5.21}$$

and

$$z = \sqrt{u_1(q)-u_2(q)} = \prod_{n=1}^{\infty}(1+q^{2n-1})^4(1-q^{2n})^2 = \left(\sum_{n=-\infty}^{\infty} q^{n^2}\right)^2. \tag{5.22}$$

Now we will apply the transformation $\tau \to -1/\tau$ to the formulas in (5.15) to deduce analogous formulas for $Q(q^4)$ and $R(q^4)$. First, replace τ with $\tau/2$, i.e., q with h, in (5.15) and (5.20), to get

$$Q(e^{i\pi\tau}) = (u_1-u_2)^2\left(1+14\left(\frac{u_3-u_2}{u_1-u_2}\right)+\left(\frac{u_3-u_2}{u_1-u_2}\right)^2\right),$$

where u_1, u_2 and u_3 represent $u_1(h)$, $u_2(h)$ and $u_3(h)$, respectively. Second, replace τ with $-1/\tau$ to get

$$Q(e^{-i\pi/\tau}) = (\underline{u}_1-\underline{u}_2)^2\left(1+14\left(\frac{\underline{u}_3-\underline{u}_2}{\underline{u}_1-\underline{u}_2}\right)+\left(\frac{\underline{u}_3-\underline{u}_2}{\underline{u}_1-\underline{u}_2}\right)^2\right).$$

Third, apply the transformation formulas (2.57) and (4.95)–(4.97) to obtain

$$(2\tau)^4 Q(e^{4\pi i\tau}) = \tau^4(u_1-u_2)^2\left(1+14\left(\frac{u_1-u_3}{u_1-u_2}\right)+\left(\frac{u_1-u_3}{u_1-u_2}\right)^2\right).$$

Finally, replace τ with 2τ and apply (5.21) and (5.22) to get

$$Q(q^4) = \frac{1}{16}z^4\left(1+14(1-x)+(1-x)^2\right) = z^4\left(1-x+\frac{x^2}{16}\right). \tag{5.23}$$

In a similar way, we may obtain the result

$$R(q^4) = z^6\left(1-\frac{x}{2}\right)\left(1-x-\frac{x^2}{32}\right). \tag{5.24}$$

5.2 Sums of eight squares and sums of eight triangular numbers

Now consider the three identities for Q given by (5.15), (5.16) and (5.23):

$$Q(q) = z^4(1 + 14x + x^2),$$
$$Q(q^2) = z^4(1 - x + x^2)$$

and

$$Q(q^4) = z^4\left(1 - x + \frac{x^2}{16}\right).$$

It follows that each of z^4, z^4x and z^4x^2 is a linear combination of $Q(q)$, $Q(q^2)$ and $Q(q^4)$, viz.,

$$z^4 = \frac{1}{15}\left(Q(q) - 2Q(q^2) + 16Q(q^4)\right),$$
$$z^4x = \frac{1}{15}\left(Q(q) - Q(q^2)\right),$$
$$z^4x^2 = \frac{16}{15}\left(Q(q^2) - Q(q^4)\right).$$

Consequently,

$$z^4(1 - x)^2 = \frac{1}{15}\left(16Q(q^2) - Q(q)\right).$$

By (5.21), (5.22) and Jacobi's triple product identity (2.21), the left hand side is

$$z^4(1-x)^2 = \prod_{n=1}^{\infty}(1 - q^{2n-1})^{16}(1 - q^{2n})^8 = \left(\sum_{n=-\infty}^{\infty}(-1)^n q^{n^2}\right)^8.$$

Therefore,

$$\left(\sum_{n=-\infty}^{\infty}(-1)^n q^{n^2}\right)^8 = 1 + 256\sum_{n=1}^{\infty}\frac{n^3q^{2n}}{1 - q^{2n}} - 16\sum_{n=1}^{\infty}\frac{n^3q^n}{1 - q^n}$$
$$= 1 + 16\sum_{n=1}^{\infty}\frac{(-1)^n n^3 q^n}{1 - q^n}.$$

This is the sum of eight squares formula. In a similar way, from (5.20), (5.22) and Gauss' formula (2.43) we have

$$z^4x = 16q\prod_{n=1}^{\infty}(1 + q^{2n-1})^8(1 + q^{2n})^8(1 - q^{2n})^8$$
$$= 16q\prod_{n=1}^{\infty}\frac{(1 - q^{2n})^8}{(1 - q^{2n-1})^8}$$
$$= 16q\left(\sum_{n=0}^{\infty} q^{n(n+1)/2}\right)^8.$$

Therefore,

$$q\left(\sum_{n=0}^{\infty} q^{n(n+1)/2}\right)^8 = \sum_{n=1}^{\infty} \frac{n^3 q^n}{1-q^n} - \sum_{n=1}^{\infty} \frac{n^3 q^{2n}}{1-q^{2n}} = \sum_{n=1}^{\infty} \frac{n^3 q^n}{1-q^{2n}}.$$

This is the sum of eight triangular numbers formula.

The corresponding results for $P(q)$, $P(q^2)$ and $P(q^4)$ will involve the derivative dz/dx, and they may be obtained as follows. From (3.65) and (5.10) we have, respectively,

$$P(q) = z^2(1-5x) + 12x(1-x)z\frac{dz}{dx} \tag{5.25}$$

and

$$2P(q^2) = P(q) + z^2(1+x).$$

Therefore,

$$P(q^2) = z^2(1-2x) + 6x(1-x)z\frac{dz}{dx}. \tag{5.26}$$

If we replace h with q in (4.34) and apply (5.22), we get

$$z^2 = u_1(q) - u_2(q) = \frac{1}{3}\left(4P(q^4) - P(q)\right),$$

so

$$P(q^4) = \frac{1}{4}\left(3z^2 + P(q)\right) = z^2\left(1 - \frac{5x}{4}\right) + 3x(1-x)z\frac{dz}{dx}. \tag{5.27}$$

5.3 Quadratic transformations

Recall the Weierstrassian invariants e_1, e_2 and e_3 were defined by (4.44) and explicit series expansions in terms of $h = e^{i\pi\tau}$ were given by (4.46)–(4.48). Let E_1, E_2 and E_3 denote the functions obtained when h is replaced with h^2 in e_1, e_2 and e_3, respectively, that is

$$E_j = E_j(q) = e_j(q) = e_j(h^2), \qquad j = 1, 2, 3. \tag{5.28}$$

The Weierstrassian invariants are related to the Halphen functions by (4.45):

$$e_j = e_j(h) = \frac{u_j(h)}{4} - \frac{P(h^2)}{12},$$

so

$$E_j = \frac{u_j(q)}{4} - \frac{P(q^2)}{12}.$$

By (5.20)–(5.22), we have

$$\begin{cases} z^2 &= u_1(q) - u_2(q) = 4(E_1 - E_2), \\ z^2 x &= u_3(q) - u_2(q) = 4(E_3 - E_2), \\ z^2(1-x) &= u_1(q) - u_3(q) = 4(E_1 - E_3). \end{cases} \tag{5.29}$$

Further, by (5.25)–(5.27), we have

$$\begin{cases} z^2 &= \frac{1}{3}\left(4P(q^4) - P(q)\right), \\ z^2 x &= -\frac{2}{3}\left(2P(q^4) - 3P(q^2) + P(q)\right), \\ z^2(1-x) &= \frac{1}{3}\left(8P(q^4) - 6P(q^2) + P(q)\right). \end{cases} \tag{5.30}$$

Combining (5.29) and (5.30) yields

$$\begin{cases} E_1 - E_2 = \frac{1}{12}\left(4P(q^4) - P(q)\right), \\ E_3 - E_2 = -\frac{1}{6}\left(2P(q^4) - 3P(q^2) + P(q)\right), \\ E_1 - E_3 = \frac{1}{12}\left(8P(q^4) - 6P(q^2) + P(q)\right). \end{cases} \tag{5.31}$$

Since $E_1 + E_2 + E_3 = 0$, we deduce that

$$\begin{cases} E_1 = \frac{1}{6}\left(2P(q^4) - P(q^2)\right), \\ E_2 = -\frac{1}{12}\left(2P(q^2) - P(q)\right), \\ E_3 = \frac{1}{12}\left(-4P(q^4) + 4P(q^2) - P(q)\right), \end{cases} \tag{5.32}$$

and from (5.29) we find that

$$\begin{cases} E_1 = \frac{z^2}{12}(2-x), \\ E_2 = -\frac{z^2}{12}(1+x), \\ E_3 = \frac{z^2}{12}(2x-1). \end{cases} \tag{5.33}$$

We will now derive the quadratic transformations for E_1, E_2 and E_3 (and hence, the quadratic transformations for e_1, e_2 and e_3) by a different method from the previous chapter. Let $\widetilde{E}_1$, $\widetilde{E}_2$ and $\widetilde{E}_3$ be defined by

$$\widetilde{E}_j = \widetilde{E}_j(q) = E_j(q^2), \qquad j = 1, 2, 3.$$

By (5.32) and (5.33),

$$\widetilde{E}_2 = -\frac{1}{2}E_1 = -\frac{z^2}{12}\left(1 - \frac{x}{2}\right) \tag{5.34}$$

and by (4.72) and (5.24),

$$\widetilde{E}_1\widetilde{E}_2\widetilde{E}_3 = \frac{R(q^4)}{864} = \frac{z^6}{864}\left(1 - \frac{x}{2}\right)\left(1 - x - \frac{x^2}{32}\right). \tag{5.35}$$

Combining (5.34) and (5.35), we get

$$\widetilde{E}_1\widetilde{E}_3 = -\frac{z^4}{72}\left(1 - x - \frac{x^2}{32}\right).$$

Now,

$$\widetilde{E}_1 + \widetilde{E}_3 = -\widetilde{E}_2 = \frac{E_1}{2} = \frac{z^2}{24}(2 - x).$$

Therefore,

$$\begin{aligned}
\left(\widetilde{E}_1 - \widetilde{E}_3\right)^2 &= \left(\widetilde{E}_1 + \widetilde{E}_3\right)^2 - 4\widetilde{E}_1\widetilde{E}_3 \\
&= \frac{z^4}{576}(4 - 4x + x^2) + \frac{z^4}{576}(32 - 32x - x^2) \\
&= \frac{z^4}{16}(1 - x) \\
&= (E_1 - E_2)(E_1 - E_3),
\end{aligned}$$

so

$$\widetilde{E}_1 - \widetilde{E}_3 = \sqrt{(E_1 - E_2)(E_1 - E_3)}. \tag{5.36}$$

The sign of the square root is made definite by comparing the first term in the expansion in powers of q on each side. From (5.34) and (5.36), and using the fact that $\widetilde{E}_1 + \widetilde{E}_2 + \widetilde{E}_3 = 0$, we obtain the quadratic transformation formulas

$$\begin{cases}
\widetilde{E}_1 = \dfrac{E_1}{4} + \dfrac{1}{2}\sqrt{(E_1 - E_2)(E_1 - E_3)}, \\[2ex]
\widetilde{E}_2 = -\dfrac{E_1}{2}, \\[2ex]
\widetilde{E}_3 = \dfrac{E_1}{4} - \dfrac{1}{2}\sqrt{(E_1 - E_2)(E_1 - E_3)},
\end{cases}$$

or, on replacing q with h,

$$\begin{cases} E_1 = \dfrac{e_1}{4} + \dfrac{1}{2}\sqrt{(e_1 - e_2)(e_1 - e_3)}, \\ E_2 = -\dfrac{e_1}{2}, \\ E_3 = \dfrac{e_1}{4} - \dfrac{1}{2}\sqrt{(e_1 - e_2)(e_1 - e_3)}, \end{cases} \tag{5.37}$$

where $e_j = e_j(h)$ and $E_j = e_j(h^2)$, $j = 1, 2, 3$. We can solve for e_j in terms of E_j as follows. Now

$$\begin{aligned} E_1 - E_2 &= \frac{3e_1}{4} + \frac{1}{2}\sqrt{(e_1 - e_2)(e_1 - e_3)} \\ &= \frac{1}{4}(e_1 - e_2) + \frac{1}{4}(e_1 - e_3) + \frac{1}{2}\sqrt{(e_1 - e_2)(e_1 - e_3)} \\ &= \frac{1}{4}\left(\sqrt{e_1 - e_2} + \sqrt{e_1 - e_3}\right)^2, \end{aligned}$$

so

$$\sqrt{e_1 - e_2} + \sqrt{e_1 - e_3} = 2\sqrt{E_1 - E_2}. \tag{5.38}$$

Similarly, we can obtain

$$\sqrt{e_1 - e_2} - \sqrt{e_1 - e_3} = 2\sqrt{E_3 - E_2}. \tag{5.39}$$

Therefore,

$$\begin{aligned} e_2 - e_1 &= \left(\sqrt{E_1 - E_2} + \sqrt{E_3 - E_2}\right)^2 \\ &= E_1 - 2E_2 + E_3 + 2\sqrt{(E_1 - E_2)(E_3 - E_2)} \\ &= -3E_2 + 2\sqrt{(E_1 - E_2)(E_3 - E_2)} \end{aligned}$$

and similarly

$$e_1 - e_3 \quad = \quad -3E_2 - 2\sqrt{(E_1 - E_2)(E_3 - E_2)}.$$

Since $e_1 + e_2 + e_3 = 0$, we deduce that

$$\begin{cases} e_1 = -2E_2, \\ e_2 = E_2 - 2\sqrt{(E_1 - E_2)(E_3 - E_2)}, \\ e_3 = E_2 + 2\sqrt{(E_1 - E_2)(E_3 - E_2)}. \end{cases} \tag{5.40}$$

The quadratic transformations for x and z may be derived from the above formulas. From (5.29) we have

$$x = \frac{E_3 - E_2}{E_1 - E_2} \qquad \text{and} \qquad z = 2\sqrt{E_1 - E_2}.$$

Now replace q with q^2 and apply (5.38) and (5.39), to get

$$x(q^2) = \frac{\widetilde{E}_3 - \widetilde{E}_2}{\widetilde{E}_1 - \widetilde{E}_2} = \left(\frac{\sqrt{E_1 - E_2} - \sqrt{E_1 - E_3}}{\sqrt{E_1 - E_2} + \sqrt{E_1 - E_3}}\right)^2 = \left(\frac{1 - \sqrt{1-x}}{1 + \sqrt{1-x}}\right)^2$$

and

$$z(q^2) = 2\sqrt{\widetilde{E}_1 - \widetilde{E}_2} = \sqrt{E_1 - E_2} + \sqrt{E_1 - E_3} = \frac{z}{2}\left(1 + \sqrt{1-x}\right).$$

5.4 The hypergeometric function ${}_2F_1(1/4, 3/4; 1; x)$

Apart from the classical connection of z and x (which are analogues of K and k, respectively, in Legendre's notation) by means of the hypergeometric function viz.,

$$z = {}_2F_1(1/2, 1/2; 1; x),$$

Ramanujan has established the connection of some transcendentals of elliptic function theory with hypergeometric functions which are different from those found in the classical development of the theory. Ramanujan gives several interesting relations in this connection specially arising in a transformation theory of these transcendentals, i.e., relations between a transcendental defined as a function of q, and the same function when q is replaced by q^n. This is entirely Ramanujan's original work, not found in existing literature. These relations are difficult to prove generally. Some of them can be proved by closely studying results (which are merely stated) found scattered in his notebooks. We shall state one of them here:

If

$$\frac{{}_2F_1(1/4, 3/4; 1; 1-\alpha)}{{}_2F_1(1/4, 3/4; 1; \alpha)} = 5\,\frac{{}_2F_1(1/4, 3/4; 1; 1-\beta)}{{}_2F_1(1/4, 3/4; 1; \beta)} \tag{5.41}$$

then

$$(\alpha\beta)^{1/2} + \{(1-\alpha)(1-\beta)\}^{1/2} \tag{5.42}$$

$$+8\{\alpha\beta(1-\alpha)(1-\beta)\}^{1/6}\left[(\alpha\beta)^{1/6} + \{(1-\alpha)(1-\beta)\}^{1/6}\right] = 1.$$

This is one of the complicated examples contained in one of the letters of Ramanujan to Hardy dated 27 February 1913, sent from Madras [24, p. 60], [94, p. 353]. It is unfortunate that the copy of these letters reprinted in Ramanujan's Collected Papers [94] contain many gaps. Two pages, numbered (8) and (10), are missing from Ramanujan's first letter to Hardy—see [24, p. 33]. We can only remark that the mathematical world would have been very grateful if they had been preserved with great care. In this connection, J. E. Littlewood, who was asked by Hardy to furnish his comments, made the remark [24, p. 69] *"I can believe that he's at least a Jacobi"* in connection with this example. We shall prove this result, and some allied ones given by Ramanujan, in Chapter 7.

We define (cf. [92, p. 260]) the functions x_2 and z_2, connected algebraically to Q and R, by

$$Q = 1 + 240 \sum_{n=1}^{\infty} \frac{n^3 q^n}{1-q^n} = z_2^4(1+3x_2) \tag{5.43}$$

and

$$R = 1 - 504 \sum_{n=1}^{\infty} \frac{n^5 q^n}{1-q^n} = z_2^6(1-9x_2). \tag{5.44}$$

We may follow the same procedure used at the end of Chapter 3. By an application of the implicit function theorem, there exist unique analytic functions $x_2 = x_2(q)$ and $z_2 = z_2(q)$ that satisfy the equations (5.43) and (5.44) and the initial conditions $x_2(0) = 0$ and $z_2(0) = 1$. Furthermore, q and z_2 may be regarded as functions of x_2 in a neighborhood of $x_2 = 0$. We will determine explicit formulas for z_2 and q in terms of x_2.

By a simple calculation we have

$$Q^3 - R^2 = 27 z_2^{12} x_2 (1-x_2)^2.$$

By the same methods used in Chapter 3, the analogues of (3.60), (3.61), (3.63), (3.64) and (3.66) are

$$P = 12\frac{z_2'}{z_2} + \left(\frac{1}{x_2} - \frac{2}{1-x_2}\right) x_2', \tag{5.45}$$

$$\frac{Q'}{Q} = \frac{4z_2'}{z_2} + \frac{3x_2'}{1+3x_2}, \tag{5.46}$$

$$x_2' = q\frac{dx_2}{dq} = z_2^2 x_2(1-x_2), \tag{5.47}$$

$$z_2' = q\frac{dz_2}{dq} = \frac{1}{12}\left(Pz_2 - (1-3x_2)z_2^3\right), \tag{5.48}$$

and

$$\frac{d}{dx_2}\left(x_2(1-x_2)\frac{dz_2}{dx_2}\right) = \frac{3z_2}{16}, \tag{5.49}$$

respectively. It follows that

$$z_2 = A\ {}_2F_1(1/4, 3/4; 1; x_2) + B\ {}_2F_1(1/4, 3/4; 1; 1-x_2)$$

for some constants A and B. To determine the values of A and B we take the limit as $x_2 \to 0$. By (3.81),

$${}_2F_1(1/4, 3/4; 1; 1-x_2) \sim -\frac{1}{\pi\sqrt{2}}\log x_2 \qquad \text{as } x_2 \to 0^+, \tag{5.50}$$

and we already know that $z_2 = 1$ when $x_2 = 0$. It follows that $A = 1$ and $B = 0$, and therefore

$$z_2 = {}_2F_1(1/4, 3/4; 1; x_2). \tag{5.51}$$

It can be checked that $z_2 \log q$ is also a solution of the differential equation (5.49). Therefore,

$${}_2F_1(1/4, 3/4; 1; x_2)\log q = C\ {}_2F_1(1/4, 3/4; 1; x_2) + D\ {}_2F_1(1/4, 3/4; 1; 1-x_2)$$

for some constants C and D, and therefore

$$\log q = C + D\frac{{}_2F_1(1/4, 3/4; 1; 1-x_2)}{{}_2F_1(1/4, 3/4; 1; x_2)}.$$

We divide both sides by $\log x_2$ and use (5.50) to compute the limit as $x_2 \to 0^+$. This gives $D = -\pi\sqrt{2}$, and it follows that

$$q = E\exp\left(-\pi\sqrt{2}\,\frac{{}_2F_1(1/4, 3/4; 1; 1-x_2)}{{}_2F_1(1/4, 3/4; 1; x_2)}\right) \tag{5.52}$$

for some constant E. In order to determine E, we divide by x_2 and take the limit as $x_2 \to 0^+$. First, by (5.43) and (5.51) we have

$$1 + 240q + O(q^2) = \left(1 + \frac{3}{16}x_2 + O(x_2^2)\right)^4 (1+3x_2) = 1 + \frac{15}{4}x_2 + O(x_2^2)$$

as $x_2 \to 0^+$. Hence,

$$\frac{q}{x_2} = \frac{1}{64}(1 + O(x_2))(1 + O(q)) \quad \text{as } x_2 \to 0^+,$$

and so

$$\lim_{x_2 \to 0^+}\frac{q}{x_2} = \frac{1}{64}. \tag{5.53}$$

On the other hand, by (3.81) and Table 3.1 we have

$$\begin{aligned}&\lim_{x_2\to 0+}\left(\pi\sqrt{2}\,{}_2F_1(1/4,3/4;1;1-x_2)+\log x_2\right)\\ &=2\Psi(1)-\Psi(1/4)-\Psi(3/4)\\ &=\log 64,\end{aligned}$$

hence we deduce from (5.52) that

$$\lim_{x_2\to 0+}\frac{q}{x_2}=\lim_{x_2\to 0+}E\exp\left(-\pi\sqrt{2}\,\frac{{}_2F_1(1/4,3/4;1;1-x_2)}{{}_2F_1(1/4,3/4;1;x_2)}\;-\;\log x_2\right)=\frac{E}{64}.$$

It follows that $E=1$, and we have proved:

$$q=\exp\left(-\pi\sqrt{2}\,\frac{{}_2F_1(1/4,3/4;1;1-x_2)}{{}_2F_1(1/2,1/2;1;x_2)}\right). \tag{5.54}$$

Ramanujan has also given in his notebooks the results [92, p. 260]:

$$Q(q^2)=z_2^4\left(1-\frac{3}{4}x_2\right)\qquad\text{and}\qquad R(q^2)=z_2^6\left(1-\frac{9}{8}x_2\right).$$

From these, it follows that

$$Q^3(q^2)-R^2(q^2)=\frac{27}{64}z_2^{12}x_2(1-x_2).$$

We apply logarithmic differentiation to get

$$2P(q^2)=12\frac{z_2'}{z_2}+\left(\frac{2}{x_2}-\frac{1}{1-x_2}\right)x_2',$$

This may be combined with (5.45) to give

$$2P(q^2)-P(q)=\frac{x'}{x(1-x)}=z_2^2,$$

where (5.47) has been used to obtain the last step. This last result viz.,

$$z_2^2=\left({}_2F_1(1/4,3/4;1;x_2)\right)^2=2P(q^2)-P(q)$$

was given by Ramanujan on p. 214 of his first notebook, but not in the second notebook [92]. Ramanujan [92, p. 260] has also given the connections of the transcendentals x_2 and z_2 with the classical functions x and z defined by (3.57) and (3.58). These are:

$$x(q)=\frac{2\sqrt{x_2(q^2)}}{1+\sqrt{x_2(q^2)}}.$$

$$z_2={}_2F_1(1/4,3/4;1;x_2)=\frac{1}{\sqrt{1+\sqrt{x_2}}}\,{}_2F_1\left(1/2,1/2;1;\frac{2\sqrt{x_2}}{1+\sqrt{x_2}}\right).$$

5.5 The hypergeometric function ${}_2F_1(1/6, 5/6; 1; x)$

We shall now investigate another hypergeometric function which Ramanujan connects with the theory of elliptic functions. First, let x_1 and z_1 be defined by

$$Q = 1 + 240 \sum_{n=1}^{\infty} \frac{n^3 q^n}{1-q^n} = z_1^4 \tag{5.55}$$

and

$$R = 1 - 504 \sum_{n=1}^{\infty} \frac{n^5 q^n}{1-q^n} = z_1^6(1-2x_1), \tag{5.56}$$

with $x_1 = 0$ and $z_1 = 1$ when $q = 0$. Then, following the same procedure as above for x_2 and z_2, and omitting most of the details, we obtain

$$Q^3 - R^2 = 4z_1^{12} x_1(1-x_1),$$

$$q\frac{dx_1}{dq} = z_1^2 x_1(1-x_1)$$

and

$$\frac{d}{dx_1}\left(x_1(1-x_1)\frac{dz_1}{dx_1}\right) = \frac{5z_1}{36}.$$

Furthermore, by (3.81) and Table 3.1 we have

$$\begin{aligned} &\lim_{x_1 \to 0+} 2\pi {}_2F_1(1/6, 5/6; 1; 1-x_1) + \log x_1 \\ &= 2\Psi(1) - \Psi(1/6) - \Psi(5/6) \\ &= \log 432. \end{aligned}$$

We may deduce that

$$z_1 = {}_2F_1(1/6, 5/6; 1; x_1) \tag{5.57}$$

and

$$q = \exp\left(-2\pi \frac{{}_2F_1(1/6, 5/6; 1; 1-x_1)}{{}_2F_1(1/6, 5/6; 1; x_1)}\right). \tag{5.58}$$

Ramanujan has also given the following transformation formula that links the hypergeometric functions for z and z_1 given by (3.69) and (5.57):

$${}_2F_1\left(\frac{1}{6}, \frac{5}{6}; 1; \frac{27}{4}\frac{p^2(1+p)^2}{(1+p+p^2)^3}\right) = \sqrt{\frac{1+p+p^2}{1+2p}}\, {}_2F_1\left(\frac{1}{2}, \frac{1}{2}; 1; \frac{p(2+p)}{1+2p}\right),$$

where $0 \le p \le 1$.

5.6 The hypergeometric function ${}_2F_1(1/3, 2/3; 1; x)$

There is one more hypergeometric function that Ramanujan connects to the theory of elliptic functions. Let x_3 and z_3 be defined by

$$Q = 1 + 240 \sum_{n=1}^{\infty} \frac{n^3 q^n}{1-q^n} = z_3^4(1+8x_3) \tag{5.59}$$

and

$$R = 1 - 504 \sum_{n=1}^{\infty} \frac{n^5 q^n}{1-q^n} = z_3^6(1 - 20x_3 - 8x_3^2), \tag{5.60}$$

with $x_3 = 0$ and $z_3 = 1$ when $q = 0$. Once again, following the same procedure as above for x_2 and z_2 and omitting most of the details, we obtain

$$Q^3 - R^2 = 64 z_3^{12} x_3 (1-x_3)^3, \tag{5.61}$$

$$q\frac{dx_3}{dq} = z_3^2 x_3(1-x_3) \tag{5.62}$$

and

$$\frac{d}{dx_3}\left(x_3(1-x_3)\frac{dz_3}{dx_3}\right) = \frac{2z_3}{9}.$$

Furthermore, by (3.81) and Table 3.1 we have

$$\begin{aligned}
&\lim_{x_3 \to 0^+} \frac{2\pi}{\sqrt{3}} {}_2F_1(1/3, 2/3; 1; 1-x_3) + \log x_3 \\
&= 2\Psi(1) - \Psi(1/3) - \Psi(2/3) \\
&= \log 27.
\end{aligned}$$

We may deduce that

$$z_3 = {}_2F_1(1/3, 2/3; 1; x_3)$$

and

$$q = \exp\left(-\frac{2\pi}{\sqrt{3}} \frac{{}_2F_1(1/3, 2/3; 1; 1-x_3)}{{}_2F_1(1/3, 2/3; 1; x_3)}\right). \tag{5.63}$$

Ramanujan has also given in his notebook the results [92, p. 260]:

$$Q(q^3) = z_3^4\left(1 - \frac{8}{9}x_3\right) \qquad \text{and} \qquad R(q^3) = z_3^6\left(1 - \frac{4}{3}x_3 + \frac{8}{27}x_3^2\right).$$

From these, it follows that

$$Q^3(q^3) - R^2(q^3) = \frac{64}{729} z_3^{12} x_3^3 (1-x_3). \tag{5.64}$$

We divide (5.64) by (5.61) and apply Jacobi's factorization of the discriminant (2.15) to get

$$q^2 \prod_{n=1}^{\infty} \frac{(1-q^{3n})^{24}}{(1-q^n)^{24}} = \frac{Q^3(q^3) - R^2(q^3)}{Q^3(q) - R^2(q)} = \frac{1}{729} \frac{x_3^2}{(1-x_3^2)}.$$

Apply logarithmic differentiation, and make use of (5.62), to obtain

$$3P(q^3) - P(q) = \left(\frac{2}{x_3} - \frac{2}{1-x_3} \right) x_3' = 2z_3^2.$$

Thus,

$$z_3^2 = \left({}_2F_1(1/3, 2/3; 1; x_3) \right)^2 = \frac{1}{2} \left(3P(q^3) - P(q) \right).$$

Ramanujan has given other interesting relations in his notebooks. For example, let x be a real number in the interval $0 < x < 1$ and define z, τ and q by

$$z = z(x) = {}_2F_1(1/3, 2/3; 1; x), \qquad \tau = \frac{i}{\sqrt{3}} \frac{z(1-x)}{z(x)} \quad \text{and} \quad q = \exp(2\pi i \tau).$$

For $0 \leq \phi \leq \pi/2$ and for a fixed value of x, define $\theta = \theta(\phi)$ by

$$\theta = \frac{1}{z} \int_0^{\phi} {}_2F_1(1/3, 2/3; 1/2; x \sin^2 t) \, dt.$$

Then,

$$\phi = \theta + 3 \sum_{n=1}^{\infty} \frac{q^n}{n(1+q^n+q^{2n})} \sin 2n\theta.$$

We note that the derivative $d\phi/d\theta$ is an interesting elliptic function of order 2 with periods π and $3\pi\tau$, and simple poles at $\theta = n\pi + m\pi\tau$, where m and n are integers and $m \not\equiv 0 \pmod 3$. Owing to lack of space we are not investigating this topic in this book.

5.7 Notes

An explanation for the deduction in (5.13) that $\widetilde{P}(q) = P(q^2)$, $\widetilde{Q}(q) = Q(q^2)$ and $\widetilde{R}(q) = R(q^2)$ is as follows. Suppose that g_1, g_2 and g_3 are functions of the form

$$g_1(q) = 1 + \sum_{n=1}^{\infty} a_n q^n, \quad g_2(q) = 1 + \sum_{n=1}^{\infty} b_n q^n, \quad \text{and} \quad g_3(q) = 1 + \sum_{n=1}^{\infty} c_n q^n \tag{5.65}$$

and that they satisfy Ramanujan's differential equations, that is,

$$12q\frac{dg_1}{dq} = g_1^2 - g_2, \quad 3q\frac{dg_2}{dq} = g_1g_2 - g_3, \quad \text{and} \quad 2q\frac{dg_3}{dq} = g_1g_3 - g_2^2. \quad (5.66)$$

If we substitute the series (5.65) into the differential equations (5.66) and equate coefficients of q, we get

$$12a_1 = 2a_1 - b_1, \qquad 3b_1 = a_1 + b_1 - c_1 \qquad \text{and} \qquad 2c_1 = a_1 + c_1 - 2b_1.$$

This system does not have a unique solution; in fact the solutions are given by

$$(a_1, b_1, c_1) = (t, -10t, 21t)$$

where t is a parameter. Ramanujan's Eisenstein series correspond to the case $t = -24$, that is,

$$(a_1, b_1, c_1) = (-24, 240, -504).$$

Now suppose $n \geq 2$, substitute the series (5.65) into the differential equations (5.66), and equate coefficients of q^n. We find that

$$\begin{aligned} 12na_n &= 2a_n - b_n + G_1(n-1), \\ 3nb_n &= a_n + b_n - c_n + G_2(n-1), \\ 2nc_n &= a_n + c_n - 2b_n + G_3(n-1), \end{aligned}$$

where $G_1(n-1)$, $G_2(n-1)$ and $G_3(n-1)$ are functions of the $3n-3$ coefficients $a_1, a_2, \ldots, a_{n-1}$, $b_1, b_2, \ldots, b_{n-1}$ and $c_1, c_2, \ldots, c_{n-1}$. Hence, a_n, b_n and c_n satisfy the linear system

$$\begin{pmatrix} 12n-2 & 1 & 0 \\ -1 & 3n-1 & 1 \\ -1 & 2 & 2n-1 \end{pmatrix} \begin{pmatrix} a_n \\ b_n \\ c_n \end{pmatrix} = \begin{pmatrix} G_1(n-1) \\ G_2(n-1) \\ G_3(n-1) \end{pmatrix}.$$

The determinant of the matrix on the left hand side is $72n^2(n-1)$. Hence, if $n \geq 2$, the coefficients a_n, b_n and c_n are determined uniquely in terms of the previous coefficients $a_1, a_2, \ldots, a_{n-1}$, $b_1, b_2, \ldots, b_{n-1}$ and $c_1, c_2, \ldots, c_{n-1}$. Therefore we have proved the following result: Suppose

(1) g_1, g_2 and g_3 are analytic in a neighborhood of $q = 0$;
(2) g_1, g_2 and g_3 satisfy Ramanujan's differential equations (5.66);
(3) $g_1(0) = g_2(0) = g_3(0) = 1$;
(4) $g_1'(0) = -24$, $g_2'(0) = 240$ and $g_3'(0) = -504$.

Table 5.1 Classification by level and signature

hypergeometric function	level	signature
${}_2F_1(1/6, 5/6; 1; x)$	1	6
${}_2F_1(1/4, 3/4; 1; x)$	2	4
${}_2F_1(1/3, 2/3; 1; x)$	3	3
${}_2F_1(1/2, 1/2; 1; x)$	4	2

Then $g_1(q) = P(q)$, $g_2(q) = Q(q)$ and $g_3(q) = R(q)$. If the fourth condition is omitted, then it is not hard to show that the solutions must be of the form $g_1(q) = P(cq)$, $g_2(q) = Q(cq)$ and $g_3(q) = R(cq)$, where c is an arbitrary constant.

The deduction in (5.13) is the application of this result with q^2 in place of q.

The topics studied in Secs. 5.4–5.6 are now collectively referred to as *Ramanujan's theories of elliptic functions to alternative bases.* Ramanujan's results appear in the unorganized pages of his second notebook [92, pp. 257–262]. It is convenient to classify the results according to their level—see Table 5.1. Sometimes a different classification according to another parameter, the signature, is used instead. Thus, Jacobi's classical results, studied in Secs. 3.4 and 4.4, belong to the theory of level 4.

For level 1, a result that is equivalent to Ramanujan's (5.58) was given by R. Fricke [54, p. 21 (26)], [56, p. 336 (10)]:

$$q = \exp\left(-\log(1728J) + \frac{G\left(\frac{1}{12}, \frac{5}{12}; 1; \frac{1}{J}\right)}{{}_2F_1\left(\frac{1}{12}, \frac{5}{12}; 1; \frac{1}{J}\right)} \right) \tag{5.67}$$

where

$$J := \frac{1}{4x_1(1-x_1)} = \frac{Q^3(q)}{Q^3(q) - R^2(q)} = \frac{1}{1728}\left(\frac{1}{q} + 744 + 196884q + \cdots\right),$$

$$G(\alpha, \beta; 1; t) := \sum_{j=1}^{\infty} \frac{(\alpha)_j(\beta)_j}{(j!)^2} \{H_j(\alpha) + H_j(\beta) - 2H_j(1)\} t^j,$$

and

$$H_j(\alpha) := \frac{1}{\alpha} + \frac{1}{\alpha+1} + \cdots + \frac{1}{\alpha+j-1}.$$

To derive Ramanujan's result (5.58) from Fricke's (or vice versa) is a routine exercise in the theory of hypergeometric functions. All of the background theory can be found in [3, Chapters 2 and 3], and the relevant identities that need to be established are

$$G\left(\frac{1}{12},\frac{5}{12};1;\frac{1}{J}\right) = (2\pi + \log(1728J))\,{}_2F_1\left(\frac{1}{12},\frac{5}{12};1;\frac{1}{J}\right) - \frac{\Gamma(\frac{1}{12})\Gamma(\frac{5}{12})}{\Gamma(\frac{1}{2})}\,{}_2F_1\left(\frac{1}{12},\frac{5}{12};\frac{1}{2};1-\frac{1}{J}\right),$$

$${}_2F_1\left(\frac{1}{12},\frac{5}{12};1;\frac{1}{J}\right) = {}_2F_1\left(\frac{1}{6},\frac{5}{6};1;x_1\right),$$

and

$${}_2F_1\left(\frac{1}{12},\frac{5}{12};\frac{1}{2};1-\frac{1}{J}\right) = \frac{\Gamma(\frac{7}{12})\Gamma(\frac{11}{12})}{2\Gamma(\frac{1}{2})}\left\{{}_2F_1\left(\frac{1}{6},\frac{5}{6};1;x_1\right) + {}_2F_1\left(\frac{1}{6},\frac{5}{6};1;1-x_1\right)\right\}.$$

The results of Fricke, e.g., (5.67) and Ramanujan, e.g., (5.58) are independent. Ramanujan's formula (5.58) should be noted for its simplicity and elegance.

The material in Secs. 5.4–5.6 of this book represents one of the first serious efforts to understand the results in pp. 257–262 of Ramanujan's second notebook. The main disadvantage of the method is that the parameterizations (5.43) and (5.44), (5.55) and (5.56), and (5.59) and (5.60) have to be known in advance and no motivation for them has been given. This deficiency has be remedied by subsequent authors, as outlined below.

Significant advances were made by J. M. Borwein and P. B. Borwein, e.g., [30], [31]; and the Borweins and F. G. Garvan, e.g., [32] and [33]. One of the major discoveries in [31] is the following beautiful cubic analogue of Jacobi's identity (4.41):

$$\left(\sum_{m=-\infty}^{\infty}\sum_{n=-\infty}^{\infty} q^{m^2+mn+n^2}\right)^3 = \left(\sum_{m=-\infty}^{\infty}\sum_{n=-\infty}^{\infty} \omega^{m-n} q^{m^2+mn+n^2}\right)^3 + \left(\sum_{m=-\infty}^{\infty}\sum_{n=-\infty}^{\infty} q^{(m+\frac{1}{3})^2+(m+\frac{1}{3})(n+\frac{1}{3})+(n+\frac{1}{3})^2}\right)^3,$$

where $\omega = \exp(2\pi i/3)$.

All of Ramanujan's results in [92, pp. 257–262] were systematically analyzed in the seminal work of Berndt, S. Bhargava and Garvan [17]. This has been reproduced and extended in [15, Chapter 33].

The theory for signature 3 was studied from a different point of view by Chan [38]. Extensions have been given by J. M. Borwein, M. D. Hirschhorn and F. G. Garvan [65], Cooper [45], R. Chapman [43] and X. M. Yang [113].

The theory for signature 4, together with applications to formulas for π, has been developed separately by Berndt, Chan and W.-C. Liaw [20].

A method that provides a uniform approach to the classical and alternative theories has been proposed in [47]. Let $r = 1, 2, 3$ or 4. Let $z_r = z_r(q)$ be defined by

$$z_r = \begin{cases} (Q(q))^{1/4} & \text{if } r = 1, \\ \left(\dfrac{rP(q^r) - P(q)}{r-1}\right)^{1/2} & \text{if } r = 2, 3 \text{ or } 4. \end{cases}$$

In the language of the theory of modular forms, z_r turns out to be a modular form of weight 1 and level r. Let $x_r = x_r(q)$ be defined to be the solution of the initial value problem

$$q\frac{dx_r}{dq} = z_r^2 x_r(1 - x_r), \qquad x_r(e^{-2\pi/\sqrt{r}}) = \frac{1}{2}.$$

It can be shown that as q increases from 0 to 1, x_r increases from 0 to 1. Therefore, the inverse function $q = q_r(x_r)$ exists. It is given explicitly by the formula

$$q = \exp\left(-\frac{2\pi}{\sqrt{r}} \frac{{}_2F_1(c_r, 1 - c_r; 1; 1 - x_r)}{{}_2F_1(c_r, 1 - c_r; 1; x_r)}\right), \tag{5.68}$$

where $c_r = 1/6$, $1/4$, $1/3$ or $1/2$, according to whether $r = 1$, 2, 3 or 4, respectively. Jacobi's formula (3.75) and Ramanujan's formulas (5.54), (5.58) and (5.63), correspond to the cases $r = 4$, 2, 1 and 3 in (5.68), respectively.

Another method, that provides a uniform approach as well as generalizations, has been developed by R. S. Maier [83].

Chapter 6

Development of Elliptic Functions

6.1 Introduction

The development of elliptic functions by Ramanujan concerns itself with the function $\wp(u)$ introduced (in a slightly heuristic way) by Eisenstein, a brilliant mathematician who unfortunately died when he was only 29 years old. Based on this work, Weierstrass built up a magnificent theory. He introduced the zeta and sigma functions, defined by

$$\zeta'(u) = -\wp(u) \quad \text{with} \quad \lim_{u\to 0}\left(\zeta(u) - \frac{1}{u}\right) = 0$$

and

$$\frac{\sigma'(u)}{\sigma(u)} = \zeta(u) \quad \text{with} \quad \lim_{u\to 0}\frac{\sigma(u)}{u} = 1.$$

Weierstrass showed that $\wp(u)$ satisfies an addition theorem and a differential equation, and that every elliptic function with the same periods as $\wp(u)$ can be expressed as a rational function of $\wp(u)$ and $\wp'(u)$. Liouville, introduced great simplification in the proofs by using Cauchy's theory of functions. Riemann, introduced the Riemann surfaces useful for the study of many-valued analytic functions, and the integrals and the inverse of one of them which were introduced and developed by the great mathematician Abel in the early years of the 19th century. Later, a related function—the elliptic modular function—was discovered and developed by Kronecker, Hermite, Dedekind, Klein, Hurwitz and others. Jacobi introduced into mathematics his famous theta functions which are connected intimately with elliptic functions. The most interesting identities of the theory are due to him (the transformation theory of elliptic functions is due mainly to Abel and Jacobi initially). The most beautiful identity of the theory is

$$\left(1+240\sum_{n=1}^{\infty}\frac{n^3q^n}{1-q^n}\right)^3-\left(1-504\sum_{n=1}^{\infty}\frac{n^5q^n}{1-q^n}\right)^2=1728q\prod_{n=1}^{\infty}(1-q^n)^{24}.$$

The usual proof is generally based on the theory of theta or sigma functions. We have seen in Chapter 2 how Ramanujan has given the simplest *a priori* proof of this identity based only on the development in power series, using his basic identity (1.5), from which all the essential fundamental theorems of elliptic function theory can be derived. Ramanujan's development has led to several interesting new results.

We have mentioned here the Eisenstein-Weierstrass theory in the preceding paragraph. Historically, the elliptic function $\wp(u)$ did not come first (as it is so in Ramanujan's treatment). The elliptic functions $\mathrm{sn}(u)$, $\mathrm{cn}(u)$ and $\mathrm{dn}(u)$ (which are denoted by $s(u)$, $c(u)$ and $\Delta(u)$, respectively, in the beautiful treatise by Hurwitz and Courant [69]) were introduced by Jacobi (1804–1851) and Abel (1802–1829). These functions arose from the classic work of Legendre [81]. Legendre's three volumes did not deal with elliptic functions as such but only with elliptic integrals, i.e., integrals of functions which are rational in x and $\sqrt{R(x)}$ where $R(x)$ is a quartic polynomial in x. The basic theorem in this connection is the addition theorem which was discovered and proved by Euler after a particular case was discovered by Fagnano a few decades earlier. Fagnano considered not only the integral $\int dx/\sqrt{1-x^4}$ but surprisingly the integral $\int dx/\sqrt{1+x^4}$ also, and he established the first theorem of the so-called complex multiplication. There are more than 50 examples of this particular case of complex multiplication in Ramanujan's notebooks. One of them is a problem proposed by Ramanujan as a question to be solved in the Journal of the Indian Mathematical Society that we met earlier in (2.59). It may be incidentally mentioned that Ramanujan's work on the general case of complex multiplication exists only as particular cases, a huge number of them. We do not know how he was able to find them working alone by himself without any knowledge of the magnificent works of Jacobi, Kronecker, Weber, Cayley and others. It shows incredible ability in dealing with expressions concerning radicals in numerical cases, as well as radicals involving rational functions, which no other previous mathematicians had treated before. Euler and Jacobi did possess the same ability but not in such profuse variety as found in Ramanujan.

Legendre's work was very extensive so far as elliptic integrals were concerned. He divided them into three kinds: first, second and third, and discovered many interesting fundamental results which were extended in an almost incredible way by Abel in his theory of integrals of algebraic functions. Legendre obtained the classic theorems of quadratic transformation and the cubic transformation (which were useful in getting approximations in applications such as the motion of the moon and the planets, etc.). He discovered the imaginary transformation of the function $P(q) = 1 - 24\sum_{n=1}^{\infty} nq^n/(1-q^n)$ in his own notation. This work was done based only on his theorem on elliptic integrals, a very difficult task. Just after the completion of this great work, Abel, and at about the same time Jacobi, simplified the theory at one stroke by considering the inverse of the elliptic integrals of the first kind. Abel unfortunately died in 1829 and Jacobi completed the work by working out the nth order transformation of elliptic functions, which gave him the clue for the parameter $q = \exp(2\pi i\tau)$. This led to his development of theta functions, particular cases of which were available in the work of Bernoulli, Euler and Gauss. Gauss also developed the theory of elliptic functions to a great extent, although this work was not published during his lifetime. The great treatise [72] published by Jacobi is in Latin, however no knowledge of Latin is necessary to understand the beautiful theorems it contains. It is unfortunate that Ramanujan did not see this book in Cambridge. The amazing formulas would have been admired by him. A particularly interesting example contained in this book is the beautiful solution of a theorem of Legendre which expresses two particular hyperelliptic integrals (involving the square root of a sextic polynomial) as the sum of two elliptic integrals each [72, p. 74]. The simple substitution employed by Jacobi in this connection would have been admired by Ramanujan and he would have discovered many such notable examples.

This long historical digression is necessary to assess the work of Ramanujan on elliptic functions which, as explained further, shows that his work was entirely independent.

Littlewood, a distinguished colleague of Hardy, has opined [62, p. 212], [82] that it may be probable that Ramanujan had seen Greenhill's book [58] on elliptic functions. But the book was available in the University of Madras Library, which was open to the public only in 1913. The library contained Cayley's book on elliptic functions [37], the comprehensive treatise in four volumes of Tannery and Molk [99] on elliptic functions (in French), the book by Appell and Lacour [5] (also in French), etc. There was a lot of red tape restrictions about borrowing books. Perhaps Ramanujan would have been allowed to borrow books from the University Library only in 1913 when he became a research scholar of the University of Madras. But in his notebooks there is no mention of double periodicity, the name "elliptic function", or the notation employed in the books on elliptic functions such as k, k', K, E, K', E', etc., of Legendre. Ramanujan uses x and z in place of k and K (where $x = k^2$ and $z = 2K/\pi$), and instead of the Weierstrassian g_2 and g_3 he uses Q and R—see (1.42) and (1.43). Although double periodicity is apparent in many of the functions introduced by Ramanujan in the notebooks, he is never conscious of it. Amazingly, he develops the imaginary transformation of several functions profusely occurring in his notebooks.

In 1913, Ramanujan became the first research scholar of the University of Madras, owing to the great efforts of Indian professors along with the effort of Sir Francis Spring, the Chairman of the Madras Port Trust. Spring was a very sincere benefactor of Ramanujan, who secured the support of the competent mathematician Professor Gilbert Walker, the head of the meteorological department of the Government of India. The English professors at that time recruited to the so-called "Indian Educational Service" directly by the Secretary of State for India in England, who knew very little mathematics, were against the grant of the scholarship. It is only the effort of an Indian Member of the Syndicate of the University of Madras, namely Justice P. R. Sundaram Iyer, who drew attention to the University of Madras Act, passed by the Parliament in London. The preamble to the act states that one of the functions of the University was to promote research. This stifled all opposition by the English professors and the first research scholarship was granted by the University of Madras to Ramanujan. In 1913, Ramanujan was busy working out certain integrals and was earnest in submitting quarterly reports on his work on mathematical research.

It is most probable that Ramanujan did not see the aforementioned books on elliptic functions. We may just quote a similar example. Hardy [62, p. 166] quotes Ramanujan's formula

$$q\frac{dP}{dq} = \frac{P^2 - Q}{12}$$

and makes the remark *"I have not seen (this result) anywhere except in Ramanujan's work"*. This was discovered by Lebesgue and Halphen in the 1860's and is contained in Halphen's famous treatise [60, p. 450]. Tannery and Molk [99] quote Halphen's earlier treatise with great admiration. Apparently Hardy had not seen Halphen's treatise in the Cambridge University Library.

Ramanujan's notebooks contain results that he had developed. He has recorded the results as such without proofs. The proofs would be in some rough work sheets which are unfortunately completely lost. In connection with the 40 identities Ramanujan gave for the Rogers-Ramanujan functions (see, e.g., [21]), Birch [28] remarked that some of these identities appear to be too complicated to guess even with his (Ramanujan's) incredible instinct for formulas. This must be true of many complicated formulas of modular functions and complex multiplication in his notebooks. It is unfortunate that his rough working of his proofs is irretrievably lost to the mathematical world.

6.2 Jacobian elliptic functions

As we have mentioned before, Ramanujan came across the Eisenstein series P, Q and R, and the differential equations which they satisfy, fairly early. He mentions these in the early part of the notebooks. Also, the differential equations (2.13), from which the Jacobi product of the discriminant modular function (2.15) is so simply proved, occur fairly early. The full proof is given by him in his published paper [91] by means of the identity (1.5):

$$\left(\frac{1}{4}\cot\frac{\theta}{2}+\sum_{n=1}^{\infty}\frac{q^n}{1-q^n}\sin n\theta\right)^2 \tag{6.1}$$

$$=\left(\frac{1}{4}\cot\frac{\theta}{2}\right)^2+\sum_{n=1}^{\infty}\frac{q^n}{(1-q^n)^2}\cos n\theta+\frac{1}{2}\sum_{n=1}^{\infty}\frac{nq^n}{1-q^n}(1-\cos n\theta).$$

The function on the left hand side of (6.1) is essentially the zeta function of Weierstrass. The notebooks contain consequences of this identity, but nowhere is the identity found in the notebooks. The identity (6.1) occurs in Jordan's book [75, p. 524] after about 150 pages of development of the theory of elliptic functions. This is the basis of Ramanujan's development of elliptic functions. We have shown earlier that the identity (6.1) and its generalized forms (1.13) and (3.17) can be used to develop the theory of the Weierstrassian elliptic function. Ramanujan came across (6.1) prior to his exposition of the Jacobian elliptic functions. We now consider his development of these functions which were defined by him as Fourier series (in terms of q-expansions) as he did with in the case of the Weierstrassian functions.

Let $y = -i\pi\tau$ so that

$$h = e^{i\pi\tau} = e^{-y}. \tag{6.2}$$

Recall from (4.23)–(4.25) and (4.45) that

$$e_1 - e_2 = \frac{1}{4}\prod_{n=1}^{\infty}(1+h^{2n-1})^8(1-h^{2n})^4,$$

$$e_1 - e_3 = \frac{1}{4}\prod_{n=1}^{\infty}(1-h^{2n-1})^8(1-h^{2n})^4,$$

and

$$e_3 - e_2 = 4h\prod_{n=1}^{\infty}(1+h^{2n})^8(1-h^{2n})^4.$$

Motivated by an entry in Ramanujan's second notebook [92, Ch. 18, Entry 14], we define three functions s, c and d, which are closely related to the Jacobian elliptic functions sn, cn and dn, respectively, by their Fourier expansions:

$$s = s(\theta) = \sum_{n=1}^{\infty}\frac{\sin(n-\frac{1}{2})\theta}{\sinh(n-\frac{1}{2})y} = 2\sum_{n=1}^{\infty}\frac{h^{n-1/2}}{1-h^{2n-1}}\sin\left(n-\frac{1}{2}\right)\theta, \tag{6.3}$$

$$c = c(\theta) = \sum_{n=1}^{\infty}\frac{\cos(n-\frac{1}{2})\theta}{\cosh(n-\frac{1}{2})y} = 2\sum_{n=1}^{\infty}\frac{h^{n-1/2}}{1+h^{2n-1}}\cos\left(n-\frac{1}{2}\right)\theta, \tag{6.4}$$

and

$$d = d(\theta) = \frac{1}{2} + \sum_{n=1}^{\infty}\frac{\cos n\theta}{\cosh ny} = \frac{1}{2} + 2\sum_{n=1}^{\infty}\frac{h^n}{1+h^{2n}}\cos n\theta. \tag{6.5}$$

Ramanujan stated results that are equivalent to

$$s^2(\theta) + c^2(\theta) = e_3 - e_2, \qquad s^2(\theta) + d^2(\theta) = e_1 - e_2, \tag{6.6}$$

and

$$s'(\theta) = c(\theta)d(\theta), \qquad c'(\theta) = -s(\theta)d(\theta), \qquad d'(\theta) = -s(\theta)c(\theta). \tag{6.7}$$

Furthermore, if ϕ is defined by

$$c(\theta) = \sqrt{e_3 - e_2}\cos\phi \qquad \text{and} \qquad s(\theta) = \sqrt{e_3 - e_2}\sin\phi,$$

then

$$\theta = \frac{1}{\sqrt{e_1 - e_2}}\int_0^{\phi}\frac{dt}{\sqrt{1-\lambda\sin^2 t}}, \qquad \text{where} \quad \lambda = \frac{e_3 - e_2}{e_1 - e_2}.$$

Differential equations for each of the functions s, c and d can be derived from the identities in (6.6) and (6.7). The addition theorem occurs as Entry 7 (viii) in Ramanujan's second notebook [92, Ch. 17] in the following interesting form, without mention of its connection with $s(\theta)$: If

$$\int_0^\alpha \frac{dt}{\sqrt{1-\lambda\sin^2 t}} + \int_0^\beta \frac{dt}{\sqrt{1-\lambda\sin^2 t}} = \int_0^\gamma \frac{dt}{\sqrt{1-\lambda\sin^2 t}},$$

then

$$\tan(\gamma/2) = \frac{\sin\alpha\sqrt{1-\lambda\sin^2\beta}+\sin\beta\sqrt{1-\lambda\sin^2\alpha}}{\cos\alpha+\cos\beta},$$

$$\gamma = \tan^{-1}\left(\tan\alpha\sqrt{1-\lambda\sin^2\beta}\right) + \tan^{-1}\left(\tan\beta\sqrt{1-\lambda\sin^2\alpha}\right),$$

$$\cot\alpha\cot\beta = \frac{\cos\gamma}{\sin\alpha\sin\beta} + \sqrt{1-\lambda\sin^2\gamma},$$

and

$$\frac{\sqrt{\lambda}}{2} = \frac{\sqrt{\sin s\sin(s-\alpha)\sin(s-\beta)\sin(s-\gamma)}}{\sin\alpha\sin\beta\sin\gamma},$$

where $2s=\alpha+\beta+\gamma$. These results will be analyzed further in Appendix C.

We shall now proceed to use Ramanujan's identity (1.5), or its equivalent form given by (1.14), to determine Fourier expansions for the squares of s, c and d. The identity (1.14), with h written in place of q, is

$$\left(\frac{1}{2}+\sum_n{}' \frac{z^n}{1-h^n}\right)^2 = \frac{1}{4}+\sum_n{}' \frac{nz^n}{1-h^n} - 2\sum_n{}' \frac{h^n z^n}{(1-h^n)^2} - \sum_n{}' \frac{h^n}{(1-h^n)^2},$$

where the primes denote that the summations are over all non-zero integers n. After some manipulations, this can be shown to be equivalent to

$$\left(\sum_n{}' \frac{z^n}{1-h^n}\right)^2 = \sum_{n=1}^\infty \frac{(n+1)(z^n+h^n z^{-n})}{1-h^n} \qquad (6.8)$$

$$-2\sum_{n=1}^\infty \frac{z^n+h^n z^{-n}}{(1-h^n)^2} - 2\sum_{n=1}^\infty \frac{nh^n}{1-h^n},$$

provided $|h|<|z|<1$. Changing z to $-z$ in (6.8) and then adding the result to (6.8), we obtain, after dividing by 2,

$$\left(\sum_n{}' \frac{z^{2n}}{1-h^{2n}}\right)^2 + \left(\sum_{n=-\infty}^\infty \frac{z^{2n-1}}{1-h^{2n-1}}\right)^2 \qquad (6.9)$$

$$= \sum_{n=1}^\infty \frac{(2n+1)(z^{2n}+h^{2n}z^{-2n})}{1-h^{2n}} - 2\sum_{n=1}^\infty \frac{z^{2n}+h^{2n}z^{-2n}}{(1-h^{2n})^2} - 2\sum_{n=1}^\infty \frac{nh^n}{1-h^n}.$$

In (6.8) we replace z with z^2 and h with h^2, to obtain

$$\left(\sum_n{}' \frac{z^{2n}}{1-h^{2n}}\right)^2 = \sum_{n=1}^{\infty} \frac{(n+1)(z^{2n}+h^{2n}z^{-2n})}{1-h^{2n}} -2\sum_{n=1}^{\infty} \frac{z^{2n}+h^{2n}z^{-2n}}{(1-h^{2n})^2} - 2\sum_{n=1}^{\infty} \frac{nh^{2n}}{1-h^{2n}}. \tag{6.10}$$

Subtracting (6.10) from (6.9) we obtain

$$\left(\sum_{n=-\infty}^{\infty} \frac{z^{2n-1}}{1-h^{2n-1}}\right)^2 = \sum_{n=1}^{\infty} \frac{n(z^{2n}+h^{2n}z^{-2n})}{1-h^{2n}} - 2\sum_{n=1}^{\infty} \frac{nh^n}{1-h^{2n}}.$$

Changing z to $h^{1/2}z$ gives

$$\left(\sum_{n=1}^{\infty} \frac{h^{(2n-1)/2}}{1-h^{2n-1}}(z^{2n-1}-z^{1-2n})\right)^2 = \sum_{n=1}^{\infty} \frac{nh^n}{1-h^{2n}}(z^{2n}+z^{-2n}) - 2\sum_{n=1}^{\infty} \frac{nh^n}{1-h^{2n}}.$$

Now put $z = \exp(i\theta/2)$ to obtain

$$\left(\sum_{n=1}^{\infty} \frac{h^{n-1/2}}{1-h^{2n-1}} \sin\left(n-\frac{1}{2}\right)\theta\right)^2 = \frac{1}{2}\sum_{n=1}^{\infty} \frac{nh^n}{1-h^{2n}}(1-\cos n\theta). \tag{6.11}$$

If we replace h with $-h$ and change θ to $\theta+\pi$ in the above, we then obtain

$$\left(\sum_{n=1}^{\infty} \frac{h^{n-1/2}}{1+h^{2n-1}} \cos\left(n-\frac{1}{2}\right)\theta\right)^2 = \frac{1}{2}\sum_{n=1}^{\infty} \frac{nh^n}{1-h^{2n}}(\cos n\theta-(-1)^n). \tag{6.12}$$

The left hand sides of (6.11) and (6.12) are essentially Fourier expansions for the squares of $s(\theta)$ and $c(\theta)$, respectively. Hermite and Jacobi obtained (6.11) and (6.12) by squaring directly the Fourier expansions of s and c. It is interesting to note that these results can also be obtained from Ramanujan's identity (1.5), or its equivalent form (1.14), as we have shown.

Before investigating properties of these functions further, we shall obtain the Fourier expansion of the square of $d(\theta)$ from Ramanujan's identity (1.5). In (6.8) first replace h with h^2, then replace z with hz to obtain

$$\left(\sum_n{}' \frac{h^n z^n}{1-h^{2n}}\right)^2 = \sum_{n=1}^{\infty} \frac{h^n}{1-h^{2n}}(n+1)(z^n+z^{-n}) -2\sum_{n=1}^{\infty} \frac{h^n}{(1-h^{2n})^2}(z^n+z^{-n}) - 2\sum_{n=1}^{\infty} \frac{nh^{2n}}{1-h^{2n}}.$$

Applying partial fractions to the left hand side, this is equivalent to

$$\left(\sum_n{}' \frac{z^n}{1-h^n} - \sum_n{}' \frac{z^n}{1+h^n}\right)^2 = 4\sum_{n=1}^{\infty} \frac{h^n}{1-h^{2n}}(n+1)(z^n+z^{-n}) \tag{6.13}$$
$$-8\sum_{n=1}^{\infty} \frac{h^n}{(1-h^{2n})^2}(z^n+z^{-n}) - 8\sum_{n=1}^{\infty} \frac{nh^{2n}}{1-h^{2n}}.$$

Next, after replacing h with h^2 in (6.8) and applying partial fractions to the left hand side, we obtain

$$\left(\sum_n{}' \frac{z^n}{1-h^n} + \sum_n{}' \frac{z^n}{1+h^n}\right)^2 \tag{6.14}$$
$$= 4\sum_{n=1}^{\infty} \frac{(n+1)(z^n+h^{2n}z^{-n})}{1-h^{2n}} - 8\sum_{n=1}^{\infty} \frac{z^n+h^{2n}z^{-n}}{(1-h^{2n})^2} - 8\sum_{n=1}^{\infty} \frac{nh^{2n}}{1-h^{2n}}.$$

Adding (6.13) and (6.14), and dividing the result by 2, we get

$$\left(\sum_n{}' \frac{z^n}{1-h^n}\right)^2 + \left(\sum_n{}' \frac{z^n}{1+h^n}\right)^2$$
$$= 2\sum_{n=1}^{\infty} \frac{(n+1)(z^n+h^nz^{-n})}{1-h^n} - 4\sum_{n=1}^{\infty} \frac{z^n+h^nz^{-n}}{(1-h^n)(1-h^{2n})} - 8\sum_{n=1}^{\infty} \frac{nh^{2n}}{1-h^{2n}}.$$

Now subtract (6.8) from this to obtain

$$\left(\sum_n{}' \frac{z^n}{1+h^n}\right)^2 = \sum_{n=1}^{\infty} \frac{(n+1)(z^n+h^nz^{-n})}{1-h^n}$$
$$-2\sum_{n=1}^{\infty} \frac{z^n+h^nz^{-n}}{1-h^{2n}} - 2\sum_{n=1}^{\infty} \frac{(-1)^n nh^n}{1-h^n}.$$

Replace h with h^2 and then change z to zh, to get

$$\left(\sum_{n=1}^{\infty} \frac{h^n}{1+h^{2n}}(z^n+z^{-n})\right)^2 = \sum_{n=1}^{\infty} \frac{(n+1)h^n}{1-h^{2n}}(z^n+z^{-n})$$
$$-2\sum_{n=1}^{\infty} \frac{h^n}{1-h^{4n}}(z^n+z^{-n}) - 2\sum_{n=1}^{\infty} \frac{(-1)^n nh^{2n}}{1-h^{2n}}.$$

Add $1/4 + \sum_{n=1}^{\infty} h^n(z^n+z^{-n})/(1+h^{2n})$ to both sides and simplify to get

$$\left(\frac{1}{2} + \sum_{n=1}^{\infty} \frac{h^n}{1+h^{2n}}(z^n+z^{-n})\right)^2$$
$$= \frac{1}{4} + \sum_{n=1}^{\infty} \frac{nh^n}{1-h^{2n}}(z^n+z^{-n}) - 2\sum_{n=1}^{\infty} \frac{(-1)^n nh^{2n}}{1-h^{2n}}.$$

Now make the substitution $z = \exp(i\theta)$ to get

$$\left(\frac{1}{4}+\sum_{n=1}^{\infty}\frac{h^n}{1+h^{2n}}\cos n\theta\right)^2 = \frac{1}{16}+\frac{1}{2}\sum_{n=1}^{\infty}\frac{nh^n}{1-h^{2n}}\cos n\theta-\frac{1}{2}\sum_{n=1}^{\infty}\frac{(-1)^n nh^{2n}}{1-h^{2n}}. \tag{6.15}$$

It may appear that the deduction of (6.15) from Ramanujan's identity (1.5) requires somewhat long computation with, however, automatic steps. Ramanujan would write the final result immediately.

The identities (6.11), (6.12) and (6.15) may be expressed in terms of the functions s, c and d, viz:

$$\begin{aligned} s^2(\theta) &= 4\left(\sum_{n=1}^{\infty}\frac{h^{n-1/2}}{1-h^{2n-1}}\sin\left(n-\frac{1}{2}\right)\theta\right)^2 \qquad (6.16)\\ &= -2\sum_{n=1}^{\infty}\frac{nh^n}{1-h^{2n}}\cos n\theta+2\sum_{n=1}^{\infty}\frac{nh^n}{1-h^{2n}}, \end{aligned}$$

$$\begin{aligned} c^2(\theta) &= 4\left(\sum_{n=1}^{\infty}\frac{h^{n-1/2}}{1+h^{2n-1}}\cos\left(n-\frac{1}{2}\right)\theta\right)^2 \qquad (6.17)\\ &= 2\sum_{n=1}^{\infty}\frac{nh^n}{1-h^{2n}}\cos n\theta-2\sum_{n=1}^{\infty}\frac{(-1)^n nh^n}{1-h^{2n}}, \end{aligned}$$

and

$$\begin{aligned} d^2(\theta) &= \left(\frac{1}{2}+2\sum_{n=1}^{\infty}\frac{h^n}{1+h^{2n}}\cos n\theta\right)^2 \qquad (6.18)\\ &= \frac{1}{4}+2\sum_{n=1}^{\infty}\frac{nh^n}{1-h^{2n}}\cos n\theta-2\sum_{n=1}^{\infty}\frac{(-1)^n nh^{2n}}{1-h^{2n}}. \end{aligned}$$

From (6.16) and (6.17), we obtain

$$\begin{aligned} s^2(\theta)+c^2(\theta) &= 4\sum_{n=1}^{\infty}\frac{(2n-1)h^{2n-1}}{1-h^{4n-2}}\\ &= 4\sum_{n=1}^{\infty}\left(\frac{nh^n}{1-h^{2n}}-\frac{2nh^{2n}}{1-h^{4n}}\right)\\ &= 4\sum_{n=1}^{\infty}\left(\frac{nh^n}{1-h^n}-\frac{nh^{2n}}{1-h^{2n}}\right)-8\sum_{n=1}^{\infty}\left(\frac{nh^{2n}}{1-h^{2n}}-\frac{nh^{4n}}{1-h^{4n}}\right)\\ &= -\frac{1}{6}\left(2P(h^4)-3P(h^2)+P(h)\right)\\ &= e_3-e_2, \end{aligned}$$

where the last step follows from (5.28) and (5.31). This proves one of the identities in (6.6); the other may be proved similarly.

By expanding as a double series and interchanging the order of summation, we may obtain the universally valid forms. For example,

$$\begin{aligned} d(\theta) &= \frac{1}{2} + \sum_{n=1}^{\infty} \frac{h^n}{1+h^{2n}}(e^{in\theta} + e^{-in\theta}) \\ &= \frac{1}{2} + \sum_{n=1}^{\infty}\sum_{m=1}^{\infty}(-1)^{m-1}h^{n(2m-1)}(e^{in\theta} + e^{-in\theta}) \\ &= \frac{1}{2} + \sum_{m=1}^{\infty}(-1)^{m-1}\left(\frac{e^{i\theta}h^{2m-1}}{1-e^{i\theta}h^{2m-1}} + \frac{e^{-i\theta}h^{2m-1}}{1-e^{-i\theta}h^{2m-1}}\right). \end{aligned} \tag{6.19}$$

This is the simplest elliptic function that can be introduced as an example in the initial teaching of function theory. One easily obtains from (6.19) the fact that $d(\theta)$ is doubly periodic with periods 2π and $4\pi\tau$.

The series for s, c and d in (6.3)–(6.5) are particular cases of the Jordan-Kronecker function, and hence can be expressed as infinite products. Using (2.75), we find that

$$s(\theta) = 2h^{1/2}\sin\frac{\theta}{2}\prod_{n=1}^{\infty}\frac{(1-e^{i\theta}h^{2n})(1-e^{-i\theta}h^{2n})(1-h^{2n})^2}{(1-e^{i\theta}h^{2n-1})(1-e^{-i\theta}h^{2n-1})(1-h^{2n-1})^2}, \tag{6.20}$$

$$c(\theta) = 2h^{1/2}\cos\frac{\theta}{2}\prod_{n=1}^{\infty}\frac{(1+e^{i\theta}h^{2n})(1+e^{-i\theta}h^{2n})(1-h^{2n})^2}{(1-e^{i\theta}h^{2n-1})(1-e^{-i\theta}h^{2n-1})(1+h^{2n-1})^2}, \tag{6.21}$$

and

$$d(\theta) = \frac{1}{2}\prod_{n=1}^{\infty}\frac{(1+e^{i\theta}h^{2n-1})(1+e^{-i\theta}h^{2n-1})(1-h^{2n})^2}{(1-e^{i\theta}h^{2n-1})(1-e^{-i\theta}h^{2n-1})(1+h^{2n})^2}. \tag{6.22}$$

The zeros, poles and periods may be deduced from the infinite products; these are recorded in Table 6.1.

We now point out that formulas for the number of representations of a positive integer as a sum of two or four squares are easily obtained using properties of the function $d(\theta)$. Take $z = 1$ in Jacobi's triple product identity (2.36) and apply Euler's identity (2.42) to get

$$\begin{aligned} \sum_{n=-\infty}^{\infty} h^{n^2} &= \prod_{n=1}^{\infty}(1+h^{2n-1})^2(1-h^{2n}) \\ &= \prod_{n=1}^{\infty}(1+h^{2n-1})(1-h^{2n}) \times \frac{(1+h^n)}{(1+h^{2n})} \\ &= \prod_{n=1}^{\infty}\frac{(1+h^{2n-1})(1-h^{2n})}{(1-h^{2n-1})(1+h^{2n})}. \end{aligned} \tag{6.23}$$

Now substitute $\theta = 0$ in (6.19) and (6.22) and equate, to get

$$\prod_{n=1}^{\infty} \frac{(1+h^{2n-1})^2(1-h^{2n})^2}{(1-h^{2n-1})^2(1+h^{2n})^2} = 1 + 4\sum_{m=1}^{\infty} \frac{(-1)^{m-1}h^{2m-1}}{1-h^{2m-1}} \tag{6.24}$$

$$= 1 + 4\sum_{j=0}^{\infty}\left(\frac{h^{4j+1}}{1-h^{4j+1}} - \frac{h^{4j+3}}{1-h^{4j+3}}\right).$$

Combining (6.23) and (6.24) we get the sum of two squares formula:

$$\left(\sum_{n=-\infty}^{\infty} h^{n^2}\right)^2 = 1 + 4\sum_{j=0}^{\infty}\left(\frac{h^{4j+1}}{1-h^{4j+1}} - \frac{h^{4j+3}}{1-h^{4j+3}}\right),$$

which we obtained earlier in (4.33) by a different method. In order to determine the number of representations of a positive integer as a sum of four squares, we observe from (6.22) that $\theta = \pi + \pi\tau$ is a zero of $d(\theta)$. We divide $d(\theta)$ by $1 + he^{-i\theta}$ and take the limit as $\theta \to \pi + \pi\tau$. Using (6.22), we get

$$\lim_{\theta\to\pi+\pi\tau} \frac{d(\theta)}{1+he^{-i\theta}} = \lim_{\theta\to\pi+\pi\tau} \frac{1}{2}\prod_{n=1}^{\infty} \frac{(1+e^{i\theta}h^{2n-1})(1+e^{-i\theta}h^{2n+1})(1-h^{2n})^2}{(1-e^{i\theta}h^{2n-1})(1-e^{-i\theta}h^{2n-1})(1+h^{2n})^2}$$

$$= \frac{1}{4}\prod_{n=1}^{\infty} \frac{(1-h^{2n})^4}{(1+h^{2n})^4}.$$

On the other hand, by (6.19) and L'Hôpital's rule we have

$$\begin{aligned}
&\lim_{\theta\to\pi+\pi\tau} \frac{d(\theta)}{1+he^{-i\theta}} \\
&= -i \lim_{\theta\to\pi+\pi\tau} d'(\theta) \\
&= -i \lim_{\theta\to\pi+\pi\tau} \sum_{m=1}^{\infty} (-1)^{m-1}\left(\frac{ie^{i\theta}h^{2m-1}}{(1-e^{i\theta}h^{2m-1})^2} - \frac{ie^{-i\theta}h^{2m-1}}{(1-e^{-i\theta}h^{2m-1})^2}\right) \\
&= \sum_{m=1}^{\infty}\left(\frac{(-1)^m h^{2m}}{(1+h^{2m})^2} - \frac{(-1)^m h^{2m-2}}{(1+h^{2m-2})^2}\right) \\
&= \frac{1}{4} + 2\sum_{m=1}^{\infty} \frac{(-1)^m h^{2m}}{(1+h^{2m})^2}.
\end{aligned}$$

Combining the last two results, we obtain

$$\prod_{n=1}^{\infty} \frac{(1-h^{2n})^4}{(1+h^{2n})^4} = 1 + 8\sum_{m=1}^{\infty} \frac{(-1)^m h^{2m}}{(1+h^{2m})^2}. \tag{6.25}$$

We now replace h^2 with $-q$. The left hand side of (6.25) is

$$\prod_{n=1}^{\infty} \frac{(1-(-q)^n)^4}{(1+(-q)^n)^4} = \prod_{n=1}^{\infty} \frac{(1+q^{2n-1})(1-q^{2n})}{(1-q^{2n-1})(1+q^{2n})} = \left(\sum_{n=-\infty}^{\infty} q^{n^2}\right)^4, \quad (6.26)$$

by (6.23). The right hand side of (6.25) becomes, on considering separately the terms when m is odd and m is even,

$$\begin{aligned} & 1+8\sum_{m=1}^{\infty} \frac{q^m}{(1+(-q)^m)^2} \qquad (6.27) \\ &= 1+8\left(\sum_{m=1}^{\infty} \frac{q^m}{(1-q^m)^2} - \sum_{m=1}^{\infty} \frac{q^{2m}}{(1-q^{2m})^2}\right) + 8\sum_{m=1}^{\infty} \frac{q^{2m}}{(1+q^{2m})^2} \\ &= 1+8\sum_{m=1}^{\infty} \frac{q^m}{(1-q^m)^2} - 32\sum_{m=1}^{\infty} \frac{q^{4m}}{(1-q^{4m})^2} \\ &= 1+8\sum_{j=1}^{\infty} \frac{jq^j}{1-q^j} - 32\sum_{j=1}^{\infty} \frac{jq^{4j}}{1-q^{4j}}, \end{aligned}$$

where the last step follows by (1.36). Combining (6.26) and (6.27) we get the sum of four squares formula:

$$\left(\sum_{n=-\infty}^{\infty} q^{n^2}\right)^4 = 1+8\sum_{j=1}^{\infty} \frac{jq^j}{1-q^j} - 32\sum_{j=1}^{\infty} \frac{jq^{4j}}{1-q^{4j}},$$

which we obtained earlier in (4.34) by a different method.

6.3 Reciprocals and quotients

If we change θ to $\theta + \pi\tau$ in the infinite product (6.20), we get, after some manipulation,

$$s(\theta+\pi\tau) = \frac{1}{s(\theta)} \times h^{1/2} \prod_{n=1}^{\infty} \frac{(1-h^{2n})^4}{(1-h^{2n-1})^4}. \quad (6.28)$$

Therefore, using (6.3) and (6.28), the Fourier expansion for the reciprocal function $1/s(\theta)$ can be determined. In a similar way, the Fourier expansions for $d(\theta)/s(\theta)$ and $c(\theta)/s(\theta)$ can be determined by changing θ to $\theta + \pi\tau$ in the functions $c(\theta)$ and $d(\theta)$, respectively.

These functions can also be studied with the help of the generalized Ramanujan identity (3.17). We recall the Jordan-Kronecker function was defined by (3.1):

$$f(\alpha, t) = \sum_{n=-\infty}^{\infty} \frac{t^n}{1-\alpha q^n},$$

and has the infinite product representation given by (3.14):

$$f(\alpha,t)=\prod_{n=1}^{\infty}\frac{(1-\alpha tq^{n-1})(1-\alpha^{-1}t^{-1}q^n)(1-q^n)^2}{(1-tq^{n-1})(1-t^{-1}q^n)(1-\alpha q^{n-1})(1-\alpha^{-1}q^n)}.$$

As always, we put

$$q=\exp(2\pi i\tau) \quad \text{and} \quad h=\exp(i\pi\tau), \quad \text{so that} \quad q=h^2.$$

Let f_1, f_2 and f_3 be defined by

$$f_1(\theta)=-if(e^{i\pi},e^{i\theta}),$$
$$f_2(\theta)=-ie^{i\theta/2}f(e^{i\pi\tau},e^{i\theta})$$

and

$$f_3(\theta)=-ie^{i\theta/2}f(e^{i\pi+i\pi\tau},e^{i\theta}).$$

Fourier series expansions may be obtained from the series definition of the Jordan-Kronecker function. For example,

$$\begin{aligned}f_1(\theta)&=-i\sum_{n=-\infty}^{\infty}\frac{e^{in\theta}}{1+h^{2n}} \qquad (6.29)\\
&=-i\left(\frac{1}{2}+\sum_{n=1}^{\infty}\frac{e^{in\theta}}{1+h^{2n}}+\sum_{n=1}^{\infty}\frac{e^{-in\theta}}{1+h^{-2n}}\right)\\
&=-i\left(\frac{1}{2}+\sum_{n=1}^{\infty}e^{in\theta}-\sum_{n=1}^{\infty}\frac{h^{2n}e^{in\theta}}{1+h^{2n}}+\sum_{n=1}^{\infty}\frac{h^{2n}e^{-in\theta}}{1+h^{2n}}\right)\\
&=-i\left(\frac{1}{2}+\frac{e^{i\theta}}{1-e^{i\theta}}-\sum_{n=1}^{\infty}\frac{h^{2n}}{1+h^{2n}}(e^{in\theta}-e^{-in\theta})\right)\\
&=\frac{1}{2}\cot\frac{\theta}{2}-2\sum_{n=1}^{\infty}\frac{h^{2n}}{1+h^{2n}}\sin n\theta.\end{aligned}$$

Similarly, we find

$$\begin{aligned}f_2(\theta)&=-ie^{i\theta/2}\sum_{n=-\infty}^{\infty}\frac{e^{in\theta}}{1-h^{2n+1}} \qquad (6.30)\\
&=\frac{1}{2}\csc\frac{\theta}{2}+2\sum_{n=0}^{\infty}\frac{h^{2n+1}}{1-h^{2n+1}}\sin\left(n+\frac{1}{2}\right)\theta,\end{aligned}$$

and

$$\begin{aligned}f_3(\theta)&=-ie^{i\theta/2}\sum_{n=-\infty}^{\infty}\frac{e^{in\theta}}{1+h^{2n+1}} \qquad (6.31)\\
&=\frac{1}{2}\csc\frac{\theta}{2}-2\sum_{n=0}^{\infty}\frac{h^{2n+1}}{1+h^{2n+1}}\sin\left(n+\frac{1}{2}\right)\theta.\end{aligned}$$

Table 6.1 Periodicity properties, zeros and poles. Here m and n are any integers.

$f(\theta)$	$\dfrac{f(\theta+2\pi m+2\pi\tau n)}{f(\theta)}$	zeros	poles
$s(\theta)$	$(-1)^m$	$\theta=2\pi m+2\pi\tau n$	$\theta=2\pi m+(2n+1)\pi\tau$
$c(\theta)$	$(-1)^{m+n}$	$\theta=(2m+1)\pi+2\pi\tau n$	$\theta=2\pi m+(2n+1)\pi\tau$
$d(\theta)$	$(-1)^n$	$\theta=(2m+1)\pi+(2n+1)\pi\tau$	$\theta=2\pi m+(2n+1)\pi\tau$
$f_1(\theta)$	$(-1)^n$	$\theta=(2m+1)\pi+2\pi\tau n$	$\theta=2\pi m+2n\pi\tau$
$f_2(\theta)$	$(-1)^m$	$\theta=2\pi m+(2n+1)\pi\tau$	$\theta=2\pi m+2n\pi\tau$
$f_3(\theta)$	$(-1)^{m+n}$	$\theta=(2m+1)\pi+(2n+1)\pi\tau$	$\theta=2\pi m+2n\pi\tau$

Infinite product formulas follow from the infinite product for the Jordan-Kronecker function. We find that

$$f_1(\theta)=\frac{1}{2}\cot\frac{\theta}{2}\prod_{n=1}^{\infty}\frac{(1+e^{i\theta}h^{2n})(1+e^{-i\theta}h^{2n})(1-h^{2n})^2}{(1-e^{i\theta}h^{2n})(1-e^{-i\theta}h^{2n})(1+h^{2n})^2},$$

$$f_2(\theta)=\frac{1}{2}\csc\frac{\theta}{2}\prod_{n=1}^{\infty}\frac{(1-e^{i\theta}h^{2n-1})(1-e^{-i\theta}h^{2n-1})(1-h^{2n})^2}{(1-e^{i\theta}h^{2n})(1-e^{-i\theta}h^{2n})(1-h^{2n-1})^2},$$

and

$$f_3(\theta)=\frac{1}{2}\csc\frac{\theta}{2}\prod_{n=1}^{\infty}\frac{(1+e^{i\theta}h^{2n-1})(1+e^{-i\theta}h^{2n-1})(1-h^{2n})^2}{(1-e^{i\theta}h^{2n})(1-e^{-i\theta}h^{2n})(1+h^{2n-1})^2}.$$

The functions f_1, f_2 and f_3 are elliptic functions. Their periods, poles and zeros can be determined from the infinite products, and these properties are recorded in Table 6.1. If we compare the infinite products for f_1, f_2 and f_3 with the products for s, c and d given by (6.20)–(6.22), we find that

$$f_1(\theta)=a_1\frac{c(\theta)}{s(\theta)},\qquad f_2(\theta)=\frac{a_2}{s(\theta)}\qquad\text{and}\qquad f_3(\theta)=a_3\frac{d(\theta)}{s(\theta)},\tag{6.32}$$

where a_1, a_2 and a_3 are independent of θ and are given by

$$a_1=\frac{1}{2}\prod_{n=1}^{\infty}\frac{(1+h^{2n-1})^2(1-h^{2n})^2}{(1-h^{2n-1})^2(1+h^{2n})^2}=\sqrt{e_1-e_2},\tag{6.33}$$

$$a_3=2h^{1/2}\prod_{n=1}^{\infty}\frac{(1+h^{2n})^2(1-h^{2n})^2}{(1+h^{2n-1})^2(1-h^{2n-1})^2}=\sqrt{e_3-e_2},\tag{6.34}$$

and

$$a_2=a_1a_3.\tag{6.35}$$

6.4 Derivatives

In the Fundamental Multiplicative Identity (3.17), let $t = e^{i\theta}$ to get

$$f(\alpha, e^{i\theta})f(\beta, e^{i\theta}) = -i\frac{\partial}{\partial\theta}f(\alpha\beta, e^{i\theta}) + f(\alpha\beta, e^{i\theta})(\rho_1(\alpha) + \rho_1(\beta)). \quad (6.36)$$

Now let $\alpha = e^{i\pi}$ and $\beta = e^{i\pi\tau}$. From (1.11) we have $\rho_1(e^{i\pi}) = 0$ and $\rho_1(e^{i\pi\tau}) = \frac{1}{2}$, hence

$$f(-1, e^{i\theta})f(h, e^{i\theta}) = -i\frac{\partial}{\partial\theta}f(-h, e^{i\theta}) + \frac{1}{2}f(-h, e^{i\theta}). \quad (6.37)$$

The left hand side of (6.37) is

$$f(-1, e^{i\theta})f(h, e^{i\theta}) = if_1(\theta)ie^{-i\theta/2}f_2(\theta) = -e^{-i\theta/2}f_1(\theta)f_2(\theta).$$

The right hand side of (6.37) is

$$\begin{aligned}
&-i\frac{\partial}{\partial\theta}\left(ie^{-i\theta/2}f_3(\theta)\right) + \frac{i}{2}e^{-i\theta/2}f_3(\theta) \\
&= e^{-i\theta/2}f_3'(\theta) - \frac{i}{2}e^{-i\theta/2}f_3(\theta) + \frac{i}{2}e^{-i\theta/2}f_3(\theta) \\
&= e^{-i\theta/2}f_3'(\theta).
\end{aligned}$$

Combining gives

$$f_3'(\theta) = -f_1(\theta)f_2(\theta). \quad (6.38)$$

Similarly, letting $\alpha = e^{i\pi\tau}$, $\beta = e^{i\pi+i\pi\tau}$ and $\alpha = e^{i\pi+i\pi\tau}$, $\beta = e^{i\pi}$ in (6.36) leads, respectively, to

$$f_1'(\theta) = -f_2(\theta)f_3(\theta), \quad (6.39)$$

$$f_2'(\theta) = -f_3(\theta)f_1(\theta). \quad (6.40)$$

Let $\beta \to 1/\alpha$ in the fundamental identity (3.17):

$$\lim_{\beta\to 1/\alpha} f(\alpha, t)f(\beta, t) = \lim_{\beta\to 1/\alpha} t\frac{\partial}{\partial t}f(\alpha\beta, t) + \lim_{\beta\to 1/\alpha} f(\alpha\beta, t)(\rho_1(\alpha) + \rho_1(\beta)). \quad (6.41)$$

The left hand side is just $f(\alpha, t)f(1/\alpha, t)$. The first limit on the right hand side is

$$\begin{aligned}
\lim_{\beta\to 1/\alpha} t\frac{\partial}{\partial t}\sum_{n=-\infty}^{\infty}\frac{t^n}{1-\alpha\beta q^n} &= \lim_{\beta\to 1/\alpha}\sum_n{}'\frac{nt^n}{1-\alpha\beta q^n} \\
&= \sum_n{}'\frac{nt^n}{1-q^n} = t\frac{d}{dt}\rho_1(t).
\end{aligned}$$

From (1.11) it follows that $\rho_1(\beta) = -\rho_1(1/\beta)$. Using this and the infinite product formula (3.14) for the function f, the remaining limit on the right hand side of equation (6.41) becomes

$$\begin{aligned}
&\lim_{\beta\to 1/\alpha} f(\alpha\beta,t)(\rho_1(\alpha)+\rho_1(\beta)) \\
&= \lim_{\beta\to 1/\alpha}(1-\alpha\beta)f(\alpha\beta,t)\lim_{\beta\to 1/\alpha}\frac{\rho_1(\alpha)+\rho_1(\beta)}{1-\alpha\beta} \\
&= -\alpha\rho_1'(\alpha).
\end{aligned}$$

Thus

$$f(\alpha,t)f(1/\alpha,t) = t\frac{d}{dt}\rho_1(t) - \alpha\frac{d}{d\alpha}\rho_1(\alpha). \tag{6.42}$$

On letting $\alpha = e^{i\gamma}$, $t = e^{i\theta}$ and using equation (1.15) and (1.28), the identity (6.42) becomes

$$f(e^{i\gamma},e^{i\theta})f(e^{-i\gamma},e^{i\theta}) = \wp(\gamma) - \wp(\theta). \tag{6.43}$$

Successively letting $\gamma = \pi$, $\gamma = \pi\tau$ and $\gamma = \pi + \pi\tau$ in (6.43), and simplifying, gives

$$f_1^2(\theta) = \wp(\theta) - e_1, \tag{6.44}$$
$$f_2^2(\theta) = \wp(\theta) - e_2, \tag{6.45}$$
$$f_3^2(\theta) = \wp(\theta) - e_3. \tag{6.46}$$

These relations show that $\wp(\theta) - e_1$, $\wp(\theta) - e_2$ and $\wp(\theta) - e_3$ are squares of elliptic functions with different periods.

The formulas (6.38)–(6.40) can be used to compute the derivatives of s, c and d. For example, using (6.32), (6.39) and (6.40), we have

$$c'(\theta) = \left(\frac{a_2 f_1(\theta)}{a_1 f_2(\theta)}\right)' = \frac{a_2}{a_1}\left(\frac{f_2(\theta)f_1'(\theta) - f_2'(\theta)f_1(\theta)}{f_2^2(\theta)}\right) = -\frac{a_2(f_2^2(\theta) - f_1^2(\theta))f_3(\theta)}{a_1 f_2^2(\theta)}.$$

Now use (6.44) and (6.45), then apply (6.32)–(6.35), to get

$$c'(\theta) = -\frac{a_2(e_1-e_2)f_3(\theta)}{a_1 f_2^2(\theta)} = -a_1^2 a_3\frac{f_3(\theta)}{f_2^2(\theta)} = -s(\theta)d(\theta).$$

This proves one of the formulas in (6.7); the others may be proved in a similar way.

6.5 Addition formulas

The Fundamental Multiplicative Identity (3.17) can be written in the form

$$f(e^{ia},e^{i\theta})f(e^{ib},e^{i\theta}) = \frac{1}{i}f(e^{i(a+b)},e^{i\theta}) + f(e^{i(a+b)},e^{i\theta})(\rho_1(e^{ia})+\rho_1(e^{ib})).$$

Apply $\partial/\partial a - \partial/\partial b$ to both sides. The result is

$$\begin{aligned}&\frac{\partial}{\partial a}f(e^{ia},e^{i\theta})f(e^{ib},e^{i\theta}) - \frac{\partial}{\partial b}f(e^{ia},e^{i\theta})f(e^{ib},e^{i\theta})\\ &= f(e^{i(a+b)},e^{i\theta})\left(\frac{d}{da}\rho_1(e^{ia}) - \frac{d}{db}\rho_1(e^{ib})\right).\end{aligned}$$

Rearranging, and using (1.15) and (1.28), gives

$$f(e^{i(a+b)},e^{i\theta}) = \frac{i\left[\frac{\partial}{\partial a}f(e^{ia},e^{i\theta})f(e^{ib},e^{i\theta}) - \frac{\partial}{\partial b}f(e^{ia},e^{i\theta})f(e^{ib},e^{i\theta})\right]}{\wp(a)-\wp(b)}. \tag{6.47}$$

Let $\theta=\pi$ to get

$$if_1(a+b) = \frac{i\left[if_1'(a)if_1(b) - if_1(a)if_1'(b)\right]}{\wp(a)-\wp(b)}.$$

Simplifying, using (6.39) and (6.44), gives

$$f_1(a+b) = \frac{f_1(a)f_2(b)f_3(b) - f_1(b)f_2(a)f_3(a)}{f_1^2(b)-f_1^2(a)}. \tag{6.48}$$

Similarly, letting $\theta=\pi\tau$ and $\theta=\pi+\pi\tau$ in (6.47) leads to

$$f_2(a+b) = \frac{f_2(a)f_3(b)f_1(b) - f_2(b)f_3(a)f_1(a)}{f_2^2(b)-f_2^2(a)} \tag{6.49}$$

and

$$f_3(a+b) = \frac{f_3(a)f_1(b)f_2(b) - f_3(b)f_1(a)f_2(a)}{f_3^2(b)-f_3^2(a)}. \tag{6.50}$$

The Jacobian elliptic functions sn, cn and dn may be defined in terms of the functions s, c and d by

$$\mathrm{sn}(u) = \frac{1}{\sqrt{e_3-e_2}}\,s\left(\frac{u}{\sqrt{e_1-e_2}}\right),$$

$$\mathrm{cn}(u) = \frac{1}{\sqrt{e_3-e_2}}\,c\left(\frac{u}{\sqrt{e_1-e_2}}\right),$$

and (note the different normalization)

$$\mathrm{dn}(u) = \frac{1}{\sqrt{e_1-e_2}}\,d\left(\frac{u}{\sqrt{e_1-e_2}}\right).$$

By (6.6), it immediately follows that

$$\operatorname{sn}^2(u) + \operatorname{cn}^2(u) = 1 \qquad \text{and} \qquad \operatorname{dn}^2(u) + k^2 \operatorname{sn}^2(u) = 1$$

where k^2 is defined by

$$k^2 = \frac{e_3 - e_2}{e_1 - e_2}.$$

From (6.7) we may readily deduce that

$$\operatorname{sn}'(\theta) = \operatorname{cn}(\theta)\operatorname{dn}(\theta),$$
$$\operatorname{cn}'(\theta) = -\operatorname{sn}(\theta)\operatorname{dn}(\theta)$$

and

$$\operatorname{dn}'(\theta) = -k^2 \operatorname{sn}(\theta)\operatorname{cn}(\theta).$$

By (6.32)–(6.35) we have

$$f_1(\theta) = \alpha\frac{\operatorname{cn}(\alpha\theta)}{\operatorname{sn}(\alpha\theta)}, \qquad f_2(\theta) = \frac{\alpha}{\operatorname{sn}(\alpha\theta)} \qquad \text{and} \qquad f_3(\theta) = \alpha\frac{\operatorname{dn}(\alpha\theta)}{\operatorname{sn}(\alpha\theta)},$$

where $\alpha = \sqrt{e_1 - e_2}$. We substitute these into (6.49) and replace a and b with a/α and b/α, respectively, to get

$$\frac{1}{\operatorname{sn}(a+b)} = \frac{\dfrac{1}{\operatorname{sn} a}\dfrac{\operatorname{dn} b}{\operatorname{sn} b}\dfrac{\operatorname{cn} b}{\operatorname{sn} b} - \dfrac{1}{\operatorname{sn} b}\dfrac{\operatorname{dn} a}{\operatorname{sn} a}\dfrac{\operatorname{cn} a}{\operatorname{sn} a}}{\dfrac{1}{\operatorname{sn}^2 b} - \dfrac{1}{\operatorname{sn}^2 a}},$$

that is,

$$\begin{aligned}
\operatorname{sn}(a+b) &= \frac{\operatorname{sn}^2 a - \operatorname{sn}^2 b}{\operatorname{sn} a \operatorname{cn} b \operatorname{dn} b - \operatorname{sn} b \operatorname{cn} a \operatorname{dn} a} \\
&= \frac{(\operatorname{sn}^2 a - \operatorname{sn}^2 b)(\operatorname{sn} a \operatorname{cn} b \operatorname{dn} b + \operatorname{sn} b \operatorname{cn} a \operatorname{dn} a)}{\operatorname{sn}^2 a \operatorname{cn}^2 b \operatorname{dn}^2 b - \operatorname{sn}^2 b \operatorname{cn}^2 a \operatorname{dn}^2 a} \\
&= \frac{(\operatorname{sn}^2 a - \operatorname{sn}^2 b)(\operatorname{sn} a \operatorname{cn} b \operatorname{dn} b + \operatorname{sn} b \operatorname{cn} a \operatorname{dn} a)}{\operatorname{sn}^2 a(1 - \operatorname{sn}^2 b)(1 - k^2 \operatorname{sn}^2 b) - \operatorname{sn}^2 b(1 - \operatorname{sn}^2 a)(1 - k^2 \operatorname{sn}^2 a)}.
\end{aligned}$$

The denominator of the last expression is seen to be

$$(\operatorname{sn}^2 a - \operatorname{sn}^2 b)(1 - k^2 \operatorname{sn}^2 a \operatorname{sn}^2 b),$$

hence we obtain the addition formula

$$\operatorname{sn}(a+b) = \frac{\operatorname{sn} a \operatorname{cn} b \operatorname{dn} b + \operatorname{sn} b \operatorname{cn} a \operatorname{dn} a}{1 - k^2 \operatorname{sn}^2 a \operatorname{sn}^2 b}.$$

Similarly, we may obtain the formulas

$$\operatorname{cn}(a+b) = \frac{\operatorname{cn} a \operatorname{cn} b - \operatorname{sn} a \operatorname{sn} b \operatorname{dn} a \operatorname{dn} b}{1 - k^2 \operatorname{sn}^2 a \operatorname{sn}^2 b}$$

and

$$\operatorname{dn}(a+b) = \frac{\operatorname{dn} a \operatorname{dn} b - k^2 \operatorname{sn} a \operatorname{sn} b \operatorname{cn} a \operatorname{cn} b}{1 - k^2 \operatorname{sn}^2 a \operatorname{sn}^2 b}.$$

In his notebooks [92, Ch. 18, Entry 18] Ramanujan has given formulas for the imaginary transformation of these functions. He has also given their Landen (quadratic) transformations.

6.6 Notes

See the paper by R. Ayoub [7] or the first chapter of the book by C. L. Siegel [98] for a description of Fagnano's work on elliptic integrals.

An investigation into the mathematics books at the University of Madras Library, as well as other books on mathematics available to Ramanujan in India, has been conducted by Berndt and R. A. Rankin [25].

A more detailed account about the awarding of the scholarship to Ramanujan has been given by S. R. Ranganathan [95, p. 30].

See [10] for information about Ramanujan's quarterly reports, and see [11] for a more detailed account.

A full and interesting account of Ramanujan's life has been given by R. Kanigel [76].

The originality and efficiency in using the fundamental multiplicative identity to develop the main properties of Jacobian elliptic functions in Secs. 6.3–6.5 should be noted. An abbreviated account of Venkatachaliengar's theory is given in the first half of [44].

Chapter 7

The Modular Function λ

7.1 Introduction

By (4.65) and (4.67), the Legendre modular function λ is given by

$$\lambda(h) = 16h \prod_{n=1}^{\infty} \frac{(1+h^{2n})^8}{(1+h^{2n-1})^8}$$

and satisfies the quadratic transformation formula

$$\lambda(h) = \frac{4\sqrt{\lambda(h^2)}}{(1+\sqrt{\lambda(h^2)})^2}. \tag{7.1}$$

In Sec. 4.3, we studied Ramanujan's observation that if the power series

$$\lambda(h) = a_1 h + a_2 h^2 + \cdots + a_n h^n + \cdots, \quad \text{where} \quad a_1 \neq 0, \tag{7.2}$$

satisfies the quadratic transformation formula (7.1), then the coefficients $a_{n+1}, a_{n+2}, \ldots, a_{2n}$ can be calculated from the first n coefficients $a_1, a_2, \ldots, a_n$. It is easy to check that if $\lambda(h)$ satisfies both (7.1) and (7.2), then $a_1 = 16$ and the coefficients a_n are all real numbers. Therefore the function $\lambda(h)$ given by the series (7.2) is characterized completely by the transformation formula (7.1). We will prove that a convergent power series of the form (7.2) and satisfying (7.1) exists in a disc of radius at least 1/16. Moreover, using only (7.1), we shall prove that

(1) $\lambda(h)$ converges in the unit disc $|h| < 1$ and has $|h| = 1$ as its natural boundary of analytic continuation;

(2) $\lambda(h) \neq 1$, $\dfrac{\lambda(h)}{h} \neq 0$ and $\lambda'(h) \neq 0$ for all $|h| < 1$.

Simple proofs, that utilize only the quadratic transformation (7.1), will be given of the modular equations of degrees 3, 5, 7, 11 and 23. We use the notation

$$\lambda_r = \lambda(h^r).$$

The modular equations we shall prove are:

$$(\lambda_1\lambda_3)^{1/4} + ((1-\lambda_1)(1-\lambda_3))^{1/4} = 1, \tag{7.3}$$

$$(\lambda_1\lambda_5)^{1/2} + ((1-\lambda_1)(1-\lambda_5))^{1/2} + \left(2^{10}\lambda_1\lambda_5(1-\lambda_1)(1-\lambda_5)\right)^{1/6} = 1, \tag{7.4}$$

$$(\lambda_1\lambda_7)^{1/8} + ((1-\lambda_1)(1-\lambda_7))^{1/8} = 1, \tag{7.5}$$

$$(\lambda_1\lambda_{11})^{1/4} + ((1-\lambda_1)(1-\lambda_{11}))^{1/4} + \left(2^{16}\lambda_1\lambda_{11}(1-\lambda_1)(1-\lambda_{11})\right)^{1/12} = 1, \tag{7.6}$$

$$(\lambda_1\lambda_{23})^{1/8} + ((1-\lambda_1)(1-\lambda_{23}))^{1/8} + \left(2^{16}\lambda_1\lambda_{23}(1-\lambda_1)(1-\lambda_{23})\right)^{1/24} = 1 \tag{7.7}$$

or, in another form given by Ramanujan,

$$1 + (\lambda_1\lambda_{23})^{1/4} + ((1-\lambda_1)(1-\lambda_{23}))^{1/4} + \left(2^{10}\lambda_1\lambda_{23}(1-\lambda_1)(1-\lambda_{23})\right)^{1/12}$$
$$= \left\{\frac{1 + (\lambda_1\lambda_{23})^{1/2} + ((1-\lambda_1)(1-\lambda_{23}))^{1/2}}{2}\right\}^{1/2}.$$

We now show that when the power series (7.2) is substituted into the functional equation (7.1), the coefficients a_n will satisfy the bound

$$|a_n| \le 16^n, \tag{7.8}$$

and therefore the power series for the function $\lambda(h)$ converges in a disc of radius at least $1/16$.

We use induction on n. If we substitute the power series (7.2) into the functional equation (7.1) and equate coefficients of h, we find that

$$a_1 = 4\sqrt{a_1}.$$

Since $a_1 \neq 0$ it follows that $a_1 = 16$. Thus (7.8) is true for $n = 1$.

Suppose that (7.8) is true for $1 \le n \le N$ for some positive integer N. We will show that (7.8) continues to hold for $N+1 \le n \le 2N$. By the binomial expansion,

$$\sqrt{\lambda(h^2)} = \left(16h^2 + a_2h^4 + a_3h^6 + \cdots + a_Nh^{2N} + O(h^{2N+2})\right)^{1/2}$$
$$= 4h\left\{1 + \frac{1}{2}\left(\frac{a_2}{16}h^2 + \frac{a_3}{16}h^4 + \cdots + \frac{a_N}{16}h^{2N-2} + \cdots\right)\right.$$
$$+ \frac{\frac{1}{2}(\frac{1}{2}-1)}{2!}\left(\frac{a_2}{16}h^2 + \frac{a_3}{16}h^4 + \cdots + \frac{a_N}{16}h^{2N-2} + \cdots\right)^2$$
$$\left.+ \frac{\frac{1}{2}(\frac{1}{2}-1)(\frac{1}{2}-2)}{3!}\left(\frac{a_2}{16}h^2 + \frac{a_3}{16}h^4 + \cdots + \frac{a_N}{16}h^{2N-2} + \cdots\right)^3 + \cdots\right\}.$$

By hypothesis, the coefficients of $h,\ h^2, \ldots, h^{2N}$ will not exceed, in absolute value, those in

$$\begin{aligned}
&4h\left\{1+\frac{1}{2}\left(16h^2+16^2h^4+\cdots+16^{N-1}h^{2N-2}+\cdots\right)\right.\\
&\qquad +\frac{1}{2}\left(16h^2+16^2h^4+\cdots+16^{N-1}h^{2N-2}+\cdots\right)^2\\
&\qquad \left.+\frac{1}{2}\left(16h^2+16^2h^4+\cdots+16^{N-1}h^{2N-2}+\cdots\right)^3+\cdots\right\}\\
&=4h\left\{1+\frac{1}{2}\left(\frac{16h^2}{1-16h^2}\right)+\frac{1}{2}\left(\frac{16h^2}{1-16h^2}\right)^2+\frac{1}{2}\left(\frac{16h^2}{1-16h^2}\right)^3+\cdots\right\}\\
&=4h\left(\frac{1-24h^2}{1-32h^2}\right),
\end{aligned}$$

and therefore the coefficients of $h,\ h^2, \ldots, h^{2N}$ in $\sqrt{\lambda(h^2)}$ will not exceed, in absolute value, those in

$$\frac{4h}{1-32h^2}.$$

It follows that the coefficients of $h,\ h^2, \ldots, h^{2N}$ in

$$\lambda(h)=\frac{4\sqrt{\lambda(h^2)}}{(1+\sqrt{\lambda(h^2)})^2}=4\sum_{n=1}^{\infty}(-1)^{n+1}n\left(\sqrt{\lambda(h^2)}\right)^n$$

are less than or equal, in absolute value, to those in

$$4\sum_{n=1}^{\infty}n\left(\frac{4h}{1-32h^2}\right)^n=\frac{16h(1-32h^2)}{(1-8h)^2(1+4h)^2},$$

whose coefficients are less than or equal to those in

$$\frac{16h}{(1-8h)^2},$$

whose coefficients in turn are less than those in

$$\frac{16h}{1-16h}=\sum_{n=1}^{\infty}(16h)^n.$$

We have shown that if (7.8) holds for $1\le n\le N$ for some positive integer N, then (7.8) holds for $1\le n\le 2N$. Therefore, by induction, the bound (7.8) has been shown to hold for all positive integers n.

The series $\lambda(h) = \sum_{n=1}^{\infty} a_n h^n = 16h + \sum_{n=2}^{\infty} a_n h^n$ converges at least in $|h| < 1/16$. Therefore, there exists a positive constant $t < 1/16$ such that in the disc $|h| < t$,

$$\begin{cases} \text{(i) } \dfrac{\lambda(h)}{h} \neq 0, \\ \text{(ii) } \lambda(h) \neq 1, \\ \text{(iii) } \lambda'(h) \neq 0, \\ \text{(iv) } \lambda(h) \text{ is analytic.} \end{cases} \tag{7.9}$$

We now show that the properties listed in (7.9) hold for all values of h in the disc $|h| < 1$. By rearranging the functional equation (7.1) we find that

$$\sqrt{1-\lambda(h)} = \frac{1-\sqrt{\lambda(h^2)}}{1+\sqrt{\lambda(h^2)}} \tag{7.10}$$

and

$$\lambda(h^2) = \left(\frac{1-\sqrt{1-\lambda(h)}}{1+\sqrt{1-\lambda(h)}}\right)^2. \tag{7.11}$$

If we apply logarithmic differentiation to (7.11), the result can be written in the form

$$h\lambda'(h)\lambda(h^2) = h^2\lambda'(h^2)\lambda(h)\sqrt{1-\lambda(h)}. \tag{7.12}$$

By (7.9)(i) we have $\lambda(h^2)/h^2 \neq 0$ for $|h^2| < t$, and by (7.9)(ii) we have $\lambda(h^2) \neq 1$ for the same range of values of h. Therefore, by (7.1) it follows that $\lambda(h)/h \neq 0$ for $|h| < t^{1/2}$. In a similar way, by (7.9)(ii) and (7.10) we may deduce that $\lambda(h) \neq 1$ in the disc $|h| < t^{1/2}$; and by (7.9)(iii) and (7.12) it follows that $\lambda'(h) \neq 0$ in the same disc. Finally, by (7.1) we deduce that $\lambda(h)$ is analytic for $|h| < t^{1/2}$. Thus, the properties (i)–(iv) in (7.9), which were known to hold in some disc $|h| < t$ for some positive number t with $t < 1/16$, continue to hold in the larger disc $|h| < t^{1/2}$. Repeating, we may extend these properties to the disc $|h| < t^{1/4}$, and so on, and we conclude that $\lambda(h)$ satisfies the properties in (7.9) in the disc $|h| < 1$.

We now show that $\lambda(h)$ cannot be continued beyond the unit disc. This will be accomplished by investigating the radial limits at various points on the unit circle.

We first claim that the radial limit as $h \to 1$ is 1, that is,

$$\lim_{h\to 1^-} \lambda(h) = 1. \tag{7.13}$$

Since the coefficients a_n are all real, it follows that $\lambda(h)$ is real-valued when h is real. Since $\lambda'(0) = 16$ and $\lambda'(h) \neq 0$ for $|h| < 1$, it follows that λ is a strictly increasing function on the real interval $0 \leq h < 1$. Moreover, $\lambda(0) = 0$. From (7.1) and (7.10) we deduce that $0 < \lambda(h) < 1$ if $0 < h < 1$. Since $\lambda(h)$ is continuous, increasing and bounded above on the interval $0 < h < 1$ it follows that the limit as $h \to 1^-$ of $\lambda(h)$ exists; let us denote the limit by ρ. Then $\rho \leq 1$. Also, for $0 < h < 1$ we have

$$\lambda(h) = \frac{4\sqrt{\lambda(h^2)}}{(1 + \sqrt{\lambda(h^2)})^2} > \sqrt{\lambda(h^2)},$$

and taking the limit as $h \to 1^-$ gives $\rho \geq \sqrt{\rho}$. It follows that $\rho = 1$ and this proves (7.13).

We now investigate the radial behavior as $h \to -1$. Since

$$\sqrt{\lambda(h^2)} = 4h \times (\text{an even function of } h),$$

the functional equation (7.1) implies that

$$\lambda(-h) = \frac{-4\sqrt{\lambda(h^2)}}{(1 - \sqrt{\lambda(h^2)})^2}. \tag{7.14}$$

Therefore, by (7.11) and (7.14) we have

$$1 - \lambda(-h) = \left(\frac{1 + \sqrt{1 - \lambda(h)}}{1 - \sqrt{1 - \lambda(h)}}\right)^2 = \frac{1}{1 - \lambda(h)},$$

so

$$\lambda(-h) = \frac{\lambda(h)}{\lambda(h) - 1}. \tag{7.15}$$

It follows that

$$\lim_{h \to -1^+} \lambda(h) = \lim_{h \to 1^-} \lambda(-h) = \lim_{h \to 1^-} \frac{\lambda(h)}{\lambda(h) - 1} = -\infty.$$

Now use (7.1) with $h^2 = q \to -1^+$ to obtain the radial limits

$$\lim_{h \to \pm \exp(i\pi/2)} \lambda(h) = \lim_{q \to -1^+} \frac{4\sqrt{\lambda(q)}}{(1 + \sqrt{\lambda(q)})^2} = 0.$$

A similar argument using (7.1) shows that for any positive integer n, every radial limit as $h \to \pm \exp(2\pi i r/2^n)$ is zero, for all integers r with $0 < r < 2^{n-1}$. Hence the unit circle $|z| = 1$ is the natural boundary and $\lambda(h)$ cannot be analytically continued beyond the unit disc.

7.2 Modular equations

If $\lambda(h)$ is the power series given by (7.2) that satisfies the quadratic transformation (7.1), and if r is any prime number, then there exists a polynomial in $\lambda(h)$ and $\lambda(h^r)$ with rational coefficients that is identically zero. The polynomial is symmetric in $\lambda(h)$ and $\lambda(h^r)$ and is invariant under the involution

$$(\lambda(h), \lambda(h^r)) \leftrightarrow (1-\lambda(h), 1-\lambda(h^r)).$$

This is a classical result about elliptic modular functions. Ramanujan, in his notebooks, has given several interesting forms of the modular equations. We do not know his methods. It is obvious that Ramanujan's work is entirely independent, because there is no mention of the term "modular equation", and the standard notation is never used by him.

We shall give a simple new method of obtaining proofs of the modular equations of degrees 3, 5, 7, 11 and 23 in a uniform way.

We define three functions $a(h)$, $b(h)$ and $c(h)$ by

$$a(h) = \frac{1}{\lambda(1-\lambda)}\, h\frac{d\lambda}{dh}, \qquad b(h) = \frac{1}{(1-\lambda)}\, h\frac{d\lambda}{dh} \quad \text{and} \quad c(h) = \frac{1}{\lambda}\, h\frac{d\lambda}{dh}. \tag{7.16}$$

For any positive integer r, let λ_r, a_r, b_r and c_r be defined by

$$\lambda_r = \lambda(h^r), \qquad a_r = a(h^r), \qquad b_r = b(h^r) \quad \text{and} \quad c_r = c(h^r).$$

Clearly $a(h) = b(h) + c(h)$, so

$$a_r = b_r + c_r. \tag{7.17}$$

It is also clear that $\lambda(h) = b(h)/a(h)$ and $1 - \lambda(h) = c(h)/a(h)$, so

$$\lambda_r = \frac{b_r}{a_r} \qquad \text{and} \qquad 1-\lambda_r = \frac{c_r}{a_r}. \tag{7.18}$$

Since $\lambda(h) = 16h - 128h^2 + O(h^3)$, the initial terms in the expansions of a, b and c are given by

$$a(h) = 1+8h+O(h^2), \qquad b(h) = 16h+O(h^2) \quad \text{and} \quad c(h) = 1-8h+O(h^2). \tag{7.19}$$

The identity (7.12) may be rewritten in the form

$$\frac{1}{\lambda_1\sqrt{1-\lambda_1}}\, h\frac{d\lambda_1}{dh} = \frac{1}{\lambda_2}\, h^2\, \frac{d\lambda_2}{dh^2}.$$

It follows that

$$\sqrt{1-\lambda_1}\, a_1 = c_2$$

and so

$$c_2 = \sqrt{a_1 c_1}.$$

By (7.1) and (7.11) we have

$$\frac{\lambda_2}{1-\lambda_2} = \frac{(1-\sqrt{1-\lambda_1})^2}{4\sqrt{1-\lambda_1}}.$$

It follows that

$$\frac{b_2}{c_2} = \frac{(\sqrt{a_1}-\sqrt{c_1})^2}{4\sqrt{a_1 c_1}}$$

and so

$$b_2 = \left(\frac{\sqrt{a_1}-\sqrt{c_1}}{2}\right)^2.$$

Lastly, since $a_r = b_r + c_r$, we obtain

$$a_2 = \left(\frac{\sqrt{a_1}+\sqrt{c_1}}{2}\right)^2.$$

In summary we have the quadratic transformation formulas:

$$a_2 = \left(\frac{\sqrt{a_1}+\sqrt{c_1}}{2}\right)^2, \qquad b_2 = \left(\frac{\sqrt{a_1}-\sqrt{c_1}}{2}\right)^2, \qquad c_2 = \sqrt{a_1 c_1}. \quad (7.20)$$

The radicals are uniquely defined by taking the first non-zero coefficient in the h-series expansion to be positive.

Let r and s be any positive integers, and let

$$\Delta_r = a_r b_r c_r. \quad (7.21)$$

The following results are simple consequences of the additive result (7.17) and the quadratic transformation formulas in (7.20) and we state them without proof:

$$\sqrt{a_{2r} b_{2r}} = \frac{b_r}{4}, \quad (7.22)$$

$$16\Delta_{2r} = b_r^{3/2}\sqrt{\Delta_r}, \quad (7.23)$$

$$\sqrt{a_{2r}a_{2s}} + \sqrt{b_{2r}b_{2s}} = \frac{1}{2}\left(\sqrt{a_r a_s} + \sqrt{c_r c_s}\right), \quad (7.24)$$

$$\sqrt{a_{2r}a_{2s}} - \sqrt{b_{2r}b_{2s}} = \frac{1}{2}\left(\sqrt{a_r c_s} + \sqrt{a_s c_r}\right), \quad (7.25)$$

$$\left(\sqrt[4]{a_{2r}a_{2s}} \pm \sqrt[4]{b_{2r}b_{2s}}\right)^2 = \frac{1}{2}\left(\sqrt{a_r a_s} \pm \sqrt{b_r b_s} + \sqrt{c_r c_s}\right), \quad (7.26)$$

$$\sqrt{a_{2r}a_{2s}} + \sqrt{b_{2r}b_{2s}} \pm \sqrt{c_{2r}c_{2s}} = \frac{1}{2}\left(\sqrt[4]{a_r a_s} \pm \sqrt[4]{c_r c_s}\right)^2. \quad (7.27)$$

The following relations can be derived using (7.20):

(R.3) If $\phi_3(h) = \left(\sqrt[4]{a_r a_s} - \sqrt[4]{b_r b_s}\right)^2 - \left(\sqrt[4]{c_r c_s}\right)^2$

then $2\phi_3(h^2) = \left(\sqrt[4]{a_r a_s} - \sqrt[4]{c_r c_s}\right)^2 - \left(\sqrt[4]{b_r b_s}\right)^2.$

(R.5) If $\phi_5(h) = \left(\sqrt{a_r a_s} - \sqrt{b_r b_s}\right)^2 - \left(\sqrt{c_r c_s} + \sqrt[6]{2^{10}\Delta_r \Delta_s}\right)^2$

then $4\phi_5(h^2) = \left(\sqrt{a_r a_s} - \sqrt{c_r c_s}\right)^2 - \left(\sqrt{b_r b_s} + \sqrt[6]{2^{10}\Delta_r \Delta_s}\right)^2.$

(R.7) If
$$\phi_7(h) = \left(\sqrt{a_r a_s} + \sqrt{b_r b_s} + \sqrt{c_r c_s} - 2\sqrt[4]{a_r a_s b_r b_s}\right)^2 - 4\left(\sqrt[4]{a_r a_s c_r c_s} + \sqrt[4]{b_r b_s c_r c_s}\right)^2$$

then
$$4\phi_7(h^2) = \left(\sqrt{a_r a_s} + \sqrt{b_r b_s} + \sqrt{c_r c_s} - 2\sqrt[4]{a_r a_s c_r c_s}\right)^2 - 4\left(\sqrt[4]{a_r a_s b_r b_s} + \sqrt[4]{b_r b_s c_r c_s}\right)^2.$$

(R.11) If $\phi_{11}(h) = \left(\sqrt[4]{a_r a_s} - \sqrt[4]{b_r b_s}\right)^2 - \left(\sqrt[4]{c_r c_s} + \sqrt[12]{2^{16}\Delta_r \Delta_s}\right)^2$

then $2\phi_{11}(h^2) = \left(\sqrt[4]{a_r a_s} - \sqrt[4]{c_r c_s}\right)^2 - \left(\sqrt[4]{b_r b_s} + \sqrt[12]{2^{16}\Delta_r \Delta_s}\right)^2.$

(R.23) If

$$\phi_{23}(h) = 4\left(\sqrt[4]{a_r a_s} + \sqrt[4]{b_r b_s}\right)^2 \times \left(\sqrt[4]{c_r c_s} + \sqrt[12]{2^{16}\Delta_r \Delta_s}\right)^2 - \left\{\left(\sqrt[4]{a_r a_s} - \sqrt[4]{b_r b_s}\right)^2 + \left(\sqrt[4]{c_r c_s} - \sqrt[12]{2^{16}\Delta_r \Delta_s}\right)^2 - 8\sqrt[6]{16\Delta_r \Delta_s}\right\}^2$$

then

$$4\phi_{23}(h^2) = 4\left(\sqrt[4]{a_r a_s} + \sqrt[4]{c_r c_s}\right)^2 \times \left(\sqrt[4]{b_r b_s} + \sqrt[12]{2^{16}\Delta_r \Delta_s}\right)^2 - \left\{\left(\sqrt[4]{a_r a_s} - \sqrt[4]{c_r c_s}\right)^2 + \left(\sqrt[4]{b_r b_s} - \sqrt[12]{2^{16}\Delta_r \Delta_s}\right)^2 - 8\sqrt[6]{16\Delta_r \Delta_s}\right\}^2.$$

We will give proofs of (R.3)–(R.23) in the next sections, and use them to derive modular equations.

7.3 Modular equation of degree 3

We begin with a proof of (R.3). If we replace h with h^2 in the definition of $\phi_3(h)$, expand and apply (7.22) and (7.27), we get

$$\begin{aligned}\phi_3(h^2) &= \left(\sqrt[4]{a_{2r}a_{2s}} - \sqrt[4]{b_{2r}b_{2s}}\right)^2 - \left(\sqrt[4]{c_{2r}c_{2s}}\right)^2 \\ &= \sqrt{a_{2r}a_{2s}} + \sqrt{b_{2r}b_{2s}} - \sqrt{c_{2r}c_{2s}} - 2\sqrt[4]{a_{2r}b_{2r}a_{2s}b_{2s}} \\ &= \frac{1}{2}\left(\sqrt[4]{a_r a_s} - \sqrt[4]{c_r c_s}\right)^2 - \frac{1}{2}\sqrt{b_r b_s},\end{aligned}$$

as required.

We will now deduce the modular equation of degree 3. Let $r = 1$ and $s = 3$. Then, we have the factorization

$$\begin{aligned}\phi_3(h) &= \left(\sqrt[4]{a_1 a_3} - \sqrt[4]{b_1 b_3} - \sqrt[4]{c_1 c_3}\right)\left(\sqrt[4]{a_1 a_3} - \sqrt[4]{b_1 b_3} + \sqrt[4]{c_1 c_3}\right) \\ &= \phi_{3,1}(h)\phi_{3,2}(h), \qquad \text{say (respectively).}\end{aligned} \tag{7.28}$$

By the identity (R.3) we also have the factorization

$$\begin{aligned}2\phi_3(h^2) &= \left(\sqrt[4]{a_1 a_3} - \sqrt[4]{b_1 b_3} - \sqrt[4]{c_1 c_3}\right)\left(\sqrt[4]{a_1 a_3} + \sqrt[4]{b_1 b_3} - \sqrt[4]{c_1 c_3}\right) \\ &= \phi_{3,1}(h)\phi_{3,3}(h), \qquad \text{say (respectively).}\end{aligned} \tag{7.29}$$

From the series expansions in (7.19), we deduce that

$$\phi_{3,1}(0) = 0, \qquad \phi_{3,2}(0) = 2, \qquad \phi_{3,3}(0) = 0 \tag{7.30}$$

and

$$\phi_{3,3}(h) = 8h + O(h^2). \tag{7.31}$$

Suppose that the expansion of $\phi_{3,1}(h)$ is given by

$$\phi_{3,1}(h) = b_s h^s + O(h^{s+1}) \tag{7.32}$$

for some positive integer s. We assume that $b_s \neq 0$ and obtain a contradiction. From (7.28), (7.30) and (7.32) we deduce

$$\phi_3(h) = \phi_{3,1}(h)\phi_{3,2}(h) = 2b_s h^s + O(h^{s+1}), \tag{7.33}$$

while from (7.29), (7.31) and (7.32) we find that

$$\phi_3(h^2) = \frac{1}{2}\phi_{3,1}(h)\phi_{3,3}(h) = 4b_s h^{s+1} + O(h^{s+2}). \tag{7.34}$$

We replace h with h^2 in (7.33) and compare with the leading coefficient in (7.34) to get

$$2b_s h^{2s} = 4b_s h^{s+1}.$$

Since $b_s \neq 0$ (by assumption), it follows that $s = 1$. But then it follows that $b_s = 0$. It follows that the expansion of $\phi_{3,1}(h)$ in (7.32) must be identically zero, and therefore we have proved

$$\sqrt[4]{a_1 a_3} = \sqrt[4]{b_1 b_3} + \sqrt[4]{c_1 c_3}.$$

If we divide by $\sqrt[4]{a_1 a_3}$ and apply (7.18) we obtain the modular equation (7.3). We can use (4.65) and (4.66) to express the modular equation (7.3) in terms of infinite products to get a beautiful identity which looks extremely difficult to prove:

$$\begin{aligned}&\prod_{n=1}^{\infty}(1+h^{2n-1})^2(1+h^{6n-3})^2\\&\quad=\prod_{n=1}^{\infty}(1-h^{2n-1})^2(1-h^{6n-3})^2+4h\prod_{n=1}^{\infty}(1+h^{2n})^2(1+h^{6n})^2.\end{aligned}$$

7.4 Modular equation of degree 5

Let us prove (R.5). Replace h with h^2 in the definition of $\phi_5(h)$ and apply (7.20), (7.23) and (7.25) to get

$$\phi_5(h^2) = \frac{1}{4}\left(\sqrt{a_r c_s} + \sqrt{a_s c_r}\right)^2 - \left(\sqrt[4]{a_r c_r a_s c_s} + 2^{1/3} b_r^{1/4} b_s^{1/4} \Delta_r^{1/12} \Delta_s^{1/12}\right)^2.$$

Multiply by 4 and expand to get

$$\begin{aligned}4\phi_5(h^2) &= a_r c_s - 2\sqrt{a_r c_r a_s c_s} + a_s c_r\\&\quad -2^{10/3}(a_r b_r c_r a_s b_s c_s)^{1/4}(\Delta_r \Delta_s)^{1/12} - 2^{8/3}(b_r b_s)^{1/2}(\Delta_r \Delta_s)^{1/6}.\end{aligned}$$

Now apply (7.17) and (7.21) and simplify, to get

$$\begin{aligned}4\phi_5(h^2) &= a_r a_s - 2\sqrt{a_r c_r a_s c_s} + c_r c_s - b_r b_s\\&\quad -2^{10/3}(\Delta_r \Delta_s)^{1/3} - 2^{8/3}(b_r b_s)^{1/2}(\Delta_r \Delta_s)^{1/6}\\&= \left(\sqrt{a_r a_s} - \sqrt{c_r c_s}\right)^2 - \left(\sqrt{b_r b_s} + \sqrt[6]{2^{10}\Delta_r \Delta_s}\right)^2.\end{aligned}$$

This proves (R.5).

We now deduce the modular equation of degree 5. Let $r = 1$ and $s = 5$. Then, we have the factorization

$$\begin{aligned}\phi_5(h) &= \left(\sqrt{a_1 a_5} - \sqrt{b_1 b_5} - \sqrt{c_1 c_5} - \sqrt[6]{2^{10}\Delta_1 \Delta_5}\right)\\&\quad \times\left(\sqrt{a_1 a_5} - \sqrt{b_1 b_5} + \sqrt{c_1 c_5} + \sqrt[6]{2^{10}\Delta_1 \Delta_5}\right)\\&= \phi_{5,1}(h)\phi_{5,2}(h), \qquad \text{say (respectively).}\end{aligned} \tag{7.35}$$

By the identity (R.5) we also have the factorization

$$\begin{aligned} 4\phi_5(h^2) &= \left(\sqrt{a_1a_5} - \sqrt{b_1b_5} - \sqrt{c_1c_5} - \sqrt[6]{2^{10}\Delta_1\Delta_5}\right) \\ &\quad \times \left(\sqrt{a_1a_5} + \sqrt{b_1b_5} - \sqrt{c_1c_5} + \sqrt[6]{2^{10}\Delta_1\Delta_5}\right) \\ &= \phi_{5,1}(h)\phi_{5,3}(h), \qquad \text{say (respectively).} \end{aligned} \tag{7.36}$$

From the series expansions in (7.19), we deduce that

$$\phi_{5,1}(0) = 0, \qquad \phi_{5,2}(0) = 2, \qquad \phi_{5,3}(0) = 0 \tag{7.37}$$

and

$$\phi_{5,3}(h) = 16h + O(h^2). \tag{7.38}$$

Suppose that the expansion of $\phi_{5,1}(h)$ is given by

$$\phi_{5,1}(h) = b_s h^s + O(h^{s+1}) \tag{7.39}$$

for some positive integer s. We assume that $b_s \neq 0$ and obtain a contradiction. From (7.35), (7.37) and (7.39) we deduce

$$\phi_5(h) = \phi_{5,1}(h)\phi_{5,2}(h) = 2b_s h^s + O(h^{s+1}), \tag{7.40}$$

while from (7.36), (7.38) and (7.39) we find that

$$\phi_5(h^2) = \frac{1}{4}\phi_{5,1}(h)\phi_{5,3}(h) = 4b_s h^{s+1} + O(h^{s+2}). \tag{7.41}$$

We replace h with h^2 in (7.40) and compare with the leading coefficient in (7.41) to get

$$2b_s h^{2s} = 4b_s h^{s+1}.$$

Since $b_s \neq 0$ (by assumption), it follows that $s = 1$. But then it follows that $b_s = 0$. It follows that the expansion of $\phi_{5,1}(h)$ in (7.39) must be identically zero, and therefore we have proved

$$\sqrt{a_1a_5} - \sqrt{b_1b_5} - \sqrt{c_1c_5} - \sqrt[6]{2^{10}\Delta_1\Delta_5} = 0.$$

If we divide by $\sqrt{a_1a_5}$ and apply (7.18) we obtain the modular equation (7.4).

We have already mentioned in (5.41) and (5.42) that in one of Ramanujan's letters to Hardy is the following result: if

$$\frac{{}_2F_1(1/4, 3/4; 1; 1-\alpha)}{{}_2F_1(1/4, 3/4; 1; \alpha)} = 5\,\frac{{}_2F_1(1/4, 3/4; 1; 1-\beta)}{{}_2F_1(1/4, 3/4; 1; \beta)} \tag{7.42}$$

then

$$(\alpha\beta)^{1/2} + \{(1-\alpha)(1-\beta)\}^{1/2} \tag{7.43}$$

$$+8\{\alpha\beta(1-\alpha)(1-\beta)\}^{1/6}\left[(\alpha\beta)^{1/6} + \{(1-\alpha)(1-\beta)\}^{1/6}\right] = 1.$$

Unless one is familiar with the work of Ramanujan in his notebooks (done independently in India before 1914) it is impossible to suggest an approach to the problem. We begin with the formulas

$$_2F_1\left(\frac{1}{2},\frac{1}{2};1;\frac{2x}{1+x}\right) = \sqrt{1+x}\,{}_2F_1\left(\frac{1}{4},\frac{3}{4};1;x^2\right) \tag{7.44}$$

and

$$_2F_1\left(\frac{1}{2},\frac{1}{2};1;\frac{1-x}{1+x}\right) = \sqrt{\frac{1+x}{2}}\,{}_2F_1\left(\frac{1}{4},\frac{3}{4};1;1-x^2\right). \tag{7.45}$$

For example, see [3, pp. 127–128, (3.17) and (3.19), resp.] We divide (7.45) by (7.44) and apply the result to (7.42) to get

$$\frac{_2F_1\left(\frac{1}{2},\frac{1}{2};1;\frac{1-\sqrt{\alpha}}{1+\sqrt{\alpha}}\right)}{_2F_1\left(\frac{1}{2},\frac{1}{2};1;\frac{2\sqrt{\alpha}}{1+\sqrt{\alpha}}\right)} = 5\,\frac{_2F_1\left(\frac{1}{2},\frac{1}{2};1;\frac{1-\sqrt{\beta}}{1+\sqrt{\beta}}\right)}{_2F_1\left(\frac{1}{2},\frac{1}{2};1;\frac{2\sqrt{\beta}}{1+\sqrt{\beta}}\right)},$$

that is,

$$\frac{_2F_1(1/2,1/2;1;1-\lambda_1)}{_2F_1(1/2,1/2;1;\lambda_1)} = 5\,\frac{_2F_1(1/2,1/2;1;1-\lambda_5)}{_2F_1(1/2,1/2;1;\lambda_5)},$$

where

$$\lambda_1 = \frac{2\sqrt{\alpha}}{1+\sqrt{\alpha}} \qquad \text{and} \qquad \lambda_5 = \frac{2\sqrt{\beta}}{1+\sqrt{\beta}}. \tag{7.46}$$

The modular equation of degree 5, (7.4), may be written in the form

$$1 - ((1-\lambda_1)(1-\lambda_5))^{1/2} = (\lambda_1\lambda_5)^{1/2} + \left(2^{10}\lambda_1\lambda_5(1-\lambda_1)(1-\lambda_5)\right)^{1/6}.$$

Therefore α and β satisfy the equation

$$\sqrt{(1+\sqrt{\alpha})(1+\sqrt{\beta})} - \sqrt{(1-\sqrt{\alpha})(1-\sqrt{\beta})}$$
$$= 2\sqrt[4]{\alpha\beta} + 4\sqrt[6]{\sqrt{\alpha\beta}(1-\alpha)(1-\beta)}.$$

If we square this result and simplify, we obtain Ramanujan's identity (7.43).

7.5 Modular equation of degree 7

Let us prove (R.7). Replace h with h^2 in the definition of $\phi_7(h)$ and apply (7.20), (7.22), (7.24) and (7.26) to get

$$\begin{aligned}\phi_7(h^2) &= \Big(\frac{1}{2}(\sqrt{a_r a_s} + \sqrt{c_r c_s}) + \sqrt[4]{a_r a_s c_r c_s} - \frac{1}{2}\sqrt{b_r b_s}\Big)^2 \\ &\qquad -2\sqrt[4]{a_r c_r a_s c_s}\left(\sqrt{a_r a_s} + \sqrt{b_r b_s} + \sqrt{c_r c_s}\right) \\ &= \frac{1}{4}\Big(\sqrt{a_r a_s} + \sqrt{b_r b_s} + \sqrt{c_r c_s} - 2\sqrt[4]{a_r c_r a_s c_s}\Big)^2 \\ &\qquad -\Big(\sqrt[4]{a_r b_r a_s b_s} + \sqrt[4]{b_r c_r b_s c_s}\Big)^2,\end{aligned}$$

where the last step follows by algebraic rearrangement. This proves (R.7).

We now deduce the modular equation of degree 7. Let $r = 1$ and $s = 7$. Then, we have the factorization

$$\begin{aligned}&\phi_7(h) \\ &= \left(\sqrt{a_1 a_7} + \sqrt{b_1 b_7} + \sqrt{c_1 c_7} - 2\sqrt[4]{a_1 a_7 b_1 b_7} - 2\sqrt[4]{a_1 a_7 c_1 c_7} - 2\sqrt[4]{b_1 b_7 c_1 c_7}\right) \\ &\quad \times\left(\sqrt{a_1 a_7} + \sqrt{b_1 b_7} + \sqrt{c_1 c_7} - 2\sqrt[4]{a_1 a_7 b_1 b_7} + 2\sqrt[4]{a_1 a_7 c_1 c_7} + 2\sqrt[4]{b_1 b_7 c_1 c_7}\right) \\ &= \phi_{7,1}(h)\phi_{7,2}(h), \qquad \text{say (respectively).}\end{aligned} \tag{7.47}$$

By the identity (R.7) we also have the factorization

$$\begin{aligned}&4\phi_7(h^2) \\ &= \left(\sqrt{a_1 a_7} + \sqrt{b_1 b_7} + \sqrt{c_1 c_7} - 2\sqrt[4]{a_1 a_7 b_1 b_7} - 2\sqrt[4]{a_1 a_7 c_1 c_7} - 2\sqrt[4]{b_1 b_7 c_1 c_7}\right) \\ &\quad \times\left(\sqrt{a_1 a_7} + \sqrt{b_1 b_7} + \sqrt{c_1 c_7} + 2\sqrt[4]{a_1 a_7 b_1 b_7} - 2\sqrt[4]{a_1 a_7 c_1 c_7} + 2\sqrt[4]{b_1 b_7 c_1 c_7}\right) \\ &= \phi_{7,1}(h)\phi_{7,3}(h), \qquad \text{say (respectively).}\end{aligned} \tag{7.48}$$

From the series expansions in (7.19), we deduce that

$$\phi_{7,1}(0) = 0, \qquad \phi_{7,2}(0) = 4, \qquad \phi_{7,3}(0) = 0 \tag{7.49}$$

and

$$\phi_{7,3}(h) = 32h^2 + O(h^3). \tag{7.50}$$

Suppose that the expansion of $\phi_{7,1}(h)$ is given by

$$\phi_{7,1}(h) = b_s h^s + O(h^{s+1}) \tag{7.51}$$

for some positive integer s. We assume that $b_s \neq 0$ and obtain a contradiction. From (7.47), (7.49) and (7.51) we deduce

$$\phi_7(h) = \phi_{7,1}(h)\phi_{7,2}(h) = 4b_s h^s + O(h^{s+1}), \tag{7.52}$$

while from (7.48), (7.50) and (7.51) we find that

$$\phi_7(h^2) = \frac{1}{4}\phi_{7,1}(h)\phi_{7,3}(h) = 8b_s h^{s+2} + O(h^{s+3}). \tag{7.53}$$

We replace h with h^2 in (7.52) and compare with the leading coefficient in (7.53) to get

$$4b_s h^{2s} = 8b_s h^{s+2}.$$

Since $b_s \neq 0$ (by assumption), it follows that $s = 2$. But then it follows that $b_s = 0$. It follows that the expansion of $\phi_{7,1}(h)$ in (7.51) must be identically zero, and therefore we have proved

$$\sqrt{a_1a_7} + \sqrt{b_1b_7} + \sqrt{c_1c_7} - 2\sqrt[4]{a_1a_7b_1b_7} - 2\sqrt[4]{a_1a_7c_1c_7} - 2\sqrt[4]{b_1b_7c_1c_7} = 0.$$

This can be resolved into factors using the identity

$$\begin{aligned} &a^4 + b^4 + c^4 - 2a^2b^2 - 2a^2c^2 - 2b^2c^2 \\ &= (a+b+c)(a+b-c)(a-b+c)(a-b-c). \end{aligned}$$

Hence $\phi_{7,1}(h)$ is a product of four factors of the form

$$\sqrt[8]{a_1a_7} \pm \sqrt[8]{b_1b_7} \pm \sqrt[8]{c_1c_7},$$

and checking the initial terms of each one of these factors it follows that we must have

$$\sqrt[8]{a_1a_7} - \sqrt[8]{b_1b_7} - \sqrt[8]{c_1c_7} = 0.$$

If we divide by $\sqrt[8]{a_1a_7}$ and apply (7.18) we obtain the modular equation (7.5). We can use (4.65) and (4.66) to express the modular equation (7.5) in terms of infinite products:

$$\begin{aligned} &\prod_{n=1}^{\infty}(1+h^{2n-1})(1+h^{14n-7}) \\ &= \prod_{n=1}^{\infty}(1-h^{2n-1})(1-h^{14n-7}) + 2h\prod_{n=1}^{\infty}(1+h^{2n})(1+h^{14n}). \end{aligned}$$

Ramanujan has given another form which is deducible from (7.5):

$$\sqrt{\frac{1+\sqrt{\lambda_1\lambda_7}+\sqrt{(1-\lambda_1)(1-\lambda_7)}}{2}} + \sqrt[8]{\lambda_1\lambda_7(1-\lambda_1)(1-\lambda_7)} = 1.$$

7.6 Modular equation of degree 11

Let us prove (R.11). Replace h with h^2 in the definition of $\phi_{11}(h)$ and apply (7.20), (7.23) and (7.26) to get

$$\phi_{11}(h^2) = \frac{1}{2}\left(\sqrt{a_r a_s} - \sqrt{b_r b_s} + \sqrt{c_r c_s}\right) - \left(\sqrt[8]{a_r a_s c_r c_s} + 2^{2/3}\sqrt[8]{b_r b_s}\sqrt[24]{\Delta_r \Delta_s}\right)^2.$$

Now multiply by 2, expand, and rearrange to get

$$\begin{aligned} 2\phi_{11}(h^2) &= \sqrt{a_r a_s} - 2\sqrt[4]{a_r a_s c_r c_s} + \sqrt{c_r c_s} \\ &\quad - \sqrt{b_r b_s} - 2^{8/3}\sqrt[6]{\Delta_r \Delta_s} - 2^{7/3}\sqrt[4]{b_r b_s}\sqrt[12]{\Delta_r \Delta_s} \\ &= \left(\sqrt[4]{a_r a_s} - \sqrt[4]{c_r c_s}\right)^2 - \left(\sqrt[4]{b_r b_s} + \sqrt[12]{2^{16}\Delta_r \Delta_s}\right)^2. \end{aligned}$$

This proves (R.11).

We now deduce the modular equation of degree 11. Let $r = 1$ and $s = 11$. Then, we have the factorization

$$\begin{aligned} \phi_{11}(h) &= \left(\sqrt[4]{a_1 a_{11}} - \sqrt[4]{b_1 b_{11}} - \sqrt[4]{c_1 c_{11}} - \sqrt[12]{2^{16}\Delta_1 \Delta_{11}}\right) \\ &\qquad \times \left(\sqrt[4]{a_1 a_{11}} - \sqrt[4]{b_1 b_{11}} + \sqrt[4]{c_1 c_{11}} + \sqrt[12]{2^{16}\Delta_1 \Delta_{11}}\right) \\ &= \phi_{11,1}(h)\phi_{11,2}(h), \qquad \text{say (respectively).} \end{aligned} \tag{7.54}$$

By the identity (R.11) we also have the factorization

$$\begin{aligned} 2\phi_{11}(h^2) &= \left(\sqrt[4]{a_1 a_{11}} - \sqrt[4]{b_1 b_{11}} - \sqrt[4]{c_1 c_{11}} - \sqrt[12]{2^{16}\Delta_1 \Delta_{11}}\right) \\ &\qquad \times \left(\sqrt[4]{a_1 a_{11}} + \sqrt[4]{b_1 b_{11}} - \sqrt[4]{c_1 c_{11}} + \sqrt[12]{2^{16}\Delta_1 \Delta_{11}}\right) \\ &= \phi_{11,1}(h)\phi_{11,3}(h), \qquad \text{say (respectively).} \end{aligned} \tag{7.55}$$

From the series expansions in (7.19), we deduce that

$$\phi_{11,1}(0) = 0, \qquad \phi_{11,2}(0) = 2, \qquad \phi_{11,3}(0) = 0 \tag{7.56}$$

and

$$\phi_{11,3}(h) = 8h + O(h^2). \tag{7.57}$$

Suppose that the expansion of $\phi_{11,1}(h)$ is given by

$$\phi_{11,1}(h) = b_s h^s + O(h^{s+1}) \tag{7.58}$$

for some positive integer s. We assume that $b_s \neq 0$ and obtain a contradiction. From (7.54), (7.56) and (7.58) we deduce

$$\phi_{11}(h) = \phi_{11,1}(h)\phi_{11,2}(h) = 2b_s h^s + O(h^{s+1}), \tag{7.59}$$

while from (7.55), (7.57) and (7.58) we find that

$$\phi_{11}(h^2) = \frac{1}{2}\phi_{11,1}(h)\phi_{11,3}(h) = 4b_s h^{s+1} + O(h^{s+2}). \tag{7.60}$$

We replace h with h^2 in (7.59) and compare with the leading coefficient in (7.60) to get

$$2b_s h^{2s} = 4b_s h^{s+1}.$$

Since $b_s \neq 0$ (by assumption), it follows that $s = 1$. But then it follows that $b_s = 0$. It follows that the expansion of $\phi_{11,1}(h)$ in (7.58) must be identically zero, and therefore we have proved

$$\sqrt[4]{a_1 a_{11}} - \sqrt[4]{b_1 b_{11}} - \sqrt[4]{c_1 c_{11}} - \sqrt[12]{2^{16}\Delta_1\Delta_{11}} = 0.$$

If we divide by $\sqrt[4]{a_1 a_{11}}$ and apply (7.18) we obtain the modular equation (7.6).

7.7 Modular equation of degree 23

Let us prove (R.23). Write

$$t = \sqrt[8]{a_r a_s}, \quad u = \sqrt[8]{b_r b_s}, \quad v = \sqrt[8]{c_r c_s}, \quad w = \sqrt[24]{2^{16}\Delta_r\Delta_s}$$

and

$$T = \sqrt[8]{a_{2r} a_{2s}}, \quad U = \sqrt[8]{b_{2r} b_{2s}}, \quad V = \sqrt[8]{c_{2r} c_{2s}}, \quad W = \sqrt[24]{2^{16}\Delta_{2r}\Delta_{2s}}.$$

Note that by (7.21), we have

$$tuv = \frac{w^3}{4} \qquad \text{and} \qquad tuvw = \sqrt[6]{16\Delta_r\Delta_s}. \tag{7.61}$$

With this notation, we have

$$\phi_{23}(h) = 4(t^2+u^2)^2(v^2+w^2)^2 - \left\{(t^2-u^2)^2 + (v^2-w^2)^2 - 8tuvw\right\}^2 \tag{7.62}$$

and

$$\phi_{23}(h^2) = 4(T^2+U^2)^2(V^2+W^2)^2 - \left\{(T^2-U^2)^2 + (V^2-W^2)^2 - 8TUVW\right\}^2. \tag{7.63}$$

By (7.20)–(7.26), we find that

$$(T^2 \pm U^2)^2 = \frac{1}{2}(t^4 \pm u^4 + v^4),$$

$$V^2 \pm W^2 = tv \pm uw$$

and

$$TUVW = \frac{u^2w^2}{4}.$$

Therefore, (7.63) becomes

$$\phi_{23}(h^2) = 2(t^4+u^4+v^4)(tv+uw)^2 - \left\{\frac{t^4-u^4+v^4}{2} + (tv-uw)^2 - 2u^2w^2\right\}^2.$$

By a tedious yet elementary calculation, this is equivalent to

$$\begin{aligned}4\phi_{23}(h^2) &= 4(t^2+v^2)^2(u^2+w^2)^2 - \left\{(t^2-v^2)^2 + (u^2-w^2)^2 - 8tuvw\right\}^2 \\ &\quad + w(4tuv - w^3)(2t^4 - 2u^4 + 2v^4 - w^4 + 4u^2w^2 + 12t^2v^2 + 12tuvw).\end{aligned}$$

This simplifies, by (7.61), to

$$4\phi_{23}(h^2) = 4(t^2+v^2)^2(u^2+w^2)^2 - \left\{(t^2-v^2)^2 + (u^2-w^2)^2 - 8tuvw\right\}^2. \tag{7.64}$$

This completes the proof of (R.23).

We will now deduce the modular equation of degree 23. From now on, let $r = 1$ and $s = 23$. From (7.62) we have the factorization

$$\phi_{23}(h) = \phi_{23,1}(h)\phi_{23,2}(h) \tag{7.65}$$

where

$$\begin{aligned}\phi_{23,1}(h) &= 2(t^2+u^2)(v^2+w^2) - (t^2-u^2)^2 - (v^2-w^2)^2 + 8tuvw \\ &= (t-u-v-w)(t-u+v+w)(t+u-v+w)(t+u+v-w)\end{aligned} \tag{7.66}$$

and

$$\phi_{23,2}(h) = 2(t^2+u^2)(v^2+w^2) + (t^2-u^2)^2 + (v^2-w^2)^2 - 8tuvw.$$

The right hand sides of (7.62) and (7.64) are the same apart from an interchange of u and v. Moreover, the function $\phi_{23,1}(h)$ in (7.66) is symmetric in u and v. It follows that the identity (7.64) factorizes as

$$4\phi_{23}(h^2) = \phi_{23,1}(h)\phi_{23,3}(h) \tag{7.67}$$

where the function $\phi_{23,3}(h)$ is obtained from $\phi_{23,2}(h)$ by interchanging s with t. The significant point is that the factorizations of $\phi_{23}(h)$ and $\phi_{23}(h^2)$ in (7.65) and (7.67) contain the common factor $\phi_{23,1}(h)$.

From the series expansions in (7.19), we deduce that

$$\phi_{23,1}(0) = 0, \qquad \phi_{23,2}(0) = 4, \qquad \phi_{23,3}(0) = 0 \tag{7.68}$$

and it can be shown that

$$\phi_{23,3}(h) = 32h^2 + O(h^3). \tag{7.69}$$

Suppose that the expansion of $\phi_{23,1}(h)$ is given by

$$\phi_{23,1}(h) = b_s h^s + O(h^{s+1}) \tag{7.70}$$

for some positive integer s. We assume that $b_s \neq 0$ and obtain a contradiction. From (7.65), (7.68) and (7.70) we deduce

$$\phi_{23}(h) = \phi_{23,1}(h)\phi_{23,2}(h) = 4b_s h^s + O(h^{s+1}), \tag{7.71}$$

while from (7.67), (7.69) and (7.70) we find that

$$\phi_{23}(h^2) = \frac{1}{4}\phi_{23,1}(h)\phi_{23,3}(h) = 8b_s h^{s+2} + O(h^{s+3}). \tag{7.72}$$

We replace h with h^2 in (7.71) and compare with the leading coefficient in (7.72) to get

$$4b_s h^{2s} = 8b_s h^{s+2}.$$

This contradicts the assumption that $b_s \neq 0$. It follows that $\phi_{23,1}(h)$ must be identically zero, and hence by (7.66) we have that the function

$$(t-u-v-w)(t-u+v+w)(t+u-v+w)(t+u+v-w)$$

is identically zero. By considering the first few terms in the expansions of each of the four factors, we deduce that

$$t-u-v-w=0,$$

identically. Therefore, we have proved

$$\sqrt[8]{a_1 a_{23}} = \sqrt[8]{b_1 b_{23}} + \sqrt[8]{c_1 c_{23}} + \sqrt[24]{2^{16}\Delta_1\Delta_{23}}.$$

This is equivalent to the modular equation (7.7).

7.8 Notes

Venkatachaliengar's method for deriving the modular equations in Secs. 7.3–7.7 is novel. The modular equations (7.3)–(7.7) appear in Ramanujan's second notebook in Chapter 19 (Entries 5(ii), 13(i) and 19(i)) and Chapter 20 (Entries 6(i) and 15(i)), respectively. The reader is referred to the corresponding sections of Berndt's book [14] for other proofs and references to original works from the 19th century. It should be emphasized that (7.3)–(7.7) represent only a tiny proportion of the modular equations in

Ramanujan's notebooks. Proofs of all of Ramanujan's modular equations have been given in the phenomenal work by Berndt [14].

The functions $a(h)$, $b(h)$ and $c(h)$ defined by (7.16) can be expressed simply in terms of Ramanujan's theta functions. By (5.14), (5.17), (5.20), (5.21) and (5.22), we have

$$\lambda(h) = x = 16h \prod_{j=1}^{\infty} \frac{(1+h^{2j})^8}{(1+h^{2j-1})^8},$$

$$1-\lambda(h) = 1-x = \prod_{j=1}^{\infty} \frac{(1-h^{2j-1})^8}{(1+h^{2j-1})^8},$$

and

$$h\frac{d\lambda}{dh} = z^2 x(1-x)$$

where

$$z = \prod_{j=1}^{\infty} (1+h^{2j-1})^4 (1-h^{2j})^2.$$

Therefore,

$$a(h) = \frac{1}{\lambda(1-\lambda)} h\frac{d\lambda}{dh} = \prod_{j=1}^{\infty} (1+h^{2j-1})^8 (1-h^{2j})^4,$$

$$b(h) = \frac{1}{(1-\lambda)} h\frac{d\lambda}{dh} = 16h \prod_{j=1}^{\infty} (1+h^{2j})^8 (1-h^{2j})^4$$

and

$$c(h) = \frac{1}{\lambda} h\frac{d\lambda}{dh} = \prod_{j=1}^{\infty} (1-h^{2j-1})^8 (1-h^{2j})^4.$$

Therefore, in terms of Ramanujan's theta functions—see (4.38) and (4.39)—we have

$$a(h) = \varphi^4(h), \qquad b(h) = 16h\psi^4(h^2) \qquad \text{and} \qquad c(h) = \varphi^4(-h).$$

The modular equations in (7.3)–(7.7) are algebraic relations between $\lambda(h)$ and $\lambda(h^n)$. We rephrase this in terms of the conventional definition of a modular equation. By (5.14),

$$\lambda(q) = x = x(q).$$

By (3.75), we have

$$q = \exp\left(-\pi \frac{{}_2F_1(1/2,1/2;1;1-x(q))}{{}_2F_1(1/2,1/2;1;x(q))}\right) \tag{7.73}$$

and if we replace q with q^n we get

$$q^n = \exp\left(-\pi \frac{{}_2F_1(1/2,1/2;1;1-x(q^n))}{{}_2F_1(1/2,1/2;1;x(q^n))}\right) \tag{7.74}$$

Comparing (7.73) and (7.74) and putting $\alpha = x(q)$, $\beta = x(q^n)$, we get

$$n \frac{{}_2F_1(1/2,1/2;1;1-\alpha)}{{}_2F_1(1/2,1/2;1;\alpha)} = \frac{{}_2F_1(1/2,1/2;1;1-\beta)}{{}_2F_1(1/2,1/2;1;\beta)}. \tag{7.75}$$

A modular equation of degree n is defined to be an algebraic relation between α and β that is implied by (7.75).

Appendix A

Singular Moduli

Apart from obtaining several forms of modular equations of various orders, Ramanujan has obtained the values of the singular moduli in terms of radicals. These arise from the modular equations when we put $t = \lambda(q) = \lambda(q^r)$ in the modular equation of degree r; for then the polynomial breaks into several factors and so each equation is solvable by radicals over the corresponding imaginary quadratic fields.

Ramanujan's paper "Modular equations and approximations to π", [90], is not properly edited and the reworking of this paper is desirable. The last two pages of Ramanujan's second notebook [92, pp. 392, 393] contain rough work with the determination of the singular moduli for several degrees, for example, 11, 35, 59, 83, 107 (in one group) and 19, 43, 67, 91, 115 and 163 (in another group). This ought to be studied in detail to find out at least some part of his method in this connection. Ramanujan recognized that the equation of the singular modulus may have degree 1. On p. 392 of his second notebook [92], Ramanujan has carried out the calculations and finds that if

$$Q = 1 + 240\sum_{n=1}^{\infty}\frac{n^3q^n}{1-q^n}, \qquad R = 1 - 504\sum_{n=1}^{\infty}\frac{n^5q^n}{1-q^n}$$

and $q = -\exp(-\pi\sqrt{D})$ where $D = 11, 19, 27, 43, 67$ or 163, then $1728Q^3/(Q^3 - R^2)$ is an integer. Ramanujan does not give any other higher integer at all in this connection, and it has now been proved that 163 is the last such integer.

We quote an extract from the lost notebook [93, p. 211]:

$$\begin{aligned}
D &= 11, & 593Q^3 - 512R^2 &= 0,\\
D &= 19, & (8^3+1)Q^3 - 8^3R^2 &= 0,\\
D &= 27, & (40^3+9)Q^3 - 40^3R^2 &= 0,\\
D &= 43, & (80^3+1)Q^3 - 80^3R^2 &= 0,\\
D &= 67, & (440^3+1)Q^3 - 440^3R^2 &= 0,\\
D &= 163, & (53360^3+1)Q^3 - 53360^3R^2 &= 0,
\end{aligned}$$

where, as before, $q = -\exp(-\pi\sqrt{D})$. Ramanujan has given the factorization into primes

$$53360^3 + 1 = 3^3 \times 7^2 \times 11^2 \times 19^2 \times 127^2 \times 163.$$

In his second notebook [92, p. 392] Ramanujan has stated the value $J_{163} = 20010$; this is equivalent to the result

$$\frac{-1728Q^3}{Q^3 - R^2} = 640320^3 = 262537412640768000$$

where $q = -\exp(-\pi\sqrt{163})$. For the case $D = 35$ Ramanujan has given the result [93, p. 211]

$$\left((60+28\sqrt{5})^3 + 27\right)Q^3 - (60+28\sqrt{5})^3R^2 = 0.$$

It is amazing how he has delved into the theory of the singular moduli, a classical topic studied by Kronecker, Weber, Klein, Fricke and others.

A.1 Notes

The results on pp. 392 and 393 of Ramanujan's second notebook [92] have been analyzed by B. C. Berndt and H. H. Chan [18]; also see Berndt's book [15, pp. 309–322]. For proofs of Ramanujan's results on p. 211 of the lost notebook [93] see the paper by Berndt and Chan [19] or the book [4, pp. 365–367].

Appendix B

The Quintuple Product Identity

Ramanujan has given in his second notebook [92, p. 202] two interesting product identities (which contain typographical errors). From them, one can obtain proofs of Ramanujan's product identities and (1.3) and (1.4) which are analogues of the classical identities (1.1) and (1.2) of Euler and Jacobi, respectively. Ramanujan's Collected Papers [94, p. 147] contain statements of the identities (1.3) and (1.4), but not proofs. We shall prove them by a simple method. These have also been rediscovered by B. Gordon [57] and others.

We start with the functions ϕ_1 and ϕ_2 defined by

$$\phi_1(z) = \sum_{n=-\infty}^{\infty} q^{(3n+1)^2} z^{3n+1} \qquad \text{and} \qquad \phi_2(z) = \sum_{n=-\infty}^{\infty} q^{(3n+2)^2} z^{3n+2}. \tag{B.1}$$

It easily follows that

$$\phi_2(z) = \phi_1(z^{-1}) \tag{B.2}$$

and

$$\phi_r(z) = q^9 z^3 \phi_r(q^6 z), \qquad r \in \{1, 2\}. \tag{B.3}$$

Let $f(z) = \phi_1(z) - \phi_2(z)$. From (B.2) and (B.3) it follows that

$$f(z) = -f(z^{-1}) \qquad \text{and} \qquad f(z) = q^9 z^3 f(q^6 z).$$

If we put $z = q^3$ we get $f(q^3) = -f(q^{-3}) = -f(q^3)$ and so $f(q^3) = 0$. In a similar way we find that $f(1) = f(-1) = 0$. Hence, f has zeros at $z = q^{6n-3}$, $\pm q^{6n}$ for all integers n (and f possibly has other zeros, too). Let

$$g(z) = (z - z^{-1}) \prod_{n=1}^{\infty} (1 - zq^{6n-3})(1 - z^{-1}q^{6n-3})(1 - z^2 q^{12n})(1 - z^{-2} q^{12n}).$$

Then g has simple zeros at $z = q^{6n-3}$, $\pm q^{6n}$, and these are the only zeros. Moreover, it is easy to check that

$$g(z) = q^9 z^3 g(q^6 z).$$

Therefore the quotient

$$h(z) = \frac{f(z)}{g(z)}$$

is analytic in $0 < |z| < \infty$, so h has a Laurent expansion given by

$$h(z) = \sum_{n=-\infty}^{\infty} c_n z^n.$$

Since h satisfies the functional equation $h(z) = h(q^6 z)$, the coefficients c_n satisfy the condition

$$(1 - q^{6n})c_n = 0,$$

and hence $h(z)$ is a constant that is independent of z. Therefore,

$$\begin{aligned} &\sum_{n=-\infty}^{\infty} q^{(3n+1)^2} \left(z^{3n+1} - z^{-3n-1}\right) \\ &= k\left(z - z^{-1}\right) \prod_{n=1}^{\infty} (1 - zq^{6n-3})(1 - z^{-1}q^{6n-3})(1 - z^2 q^{12n})(1 - z^{-2}q^{12n}), \end{aligned} \tag{B.4}$$

where k depends on q but not on z. We evaluate the constant k by putting $z = \omega := \exp(2\pi i/3)$. The left hand side of (B.4) becomes, using Jacobi's triple product identity (2.36):

$$\begin{aligned} &(\omega - \omega^{-1}) \sum_{n=-\infty}^{\infty} q^{(3n+1)^2} \\ &= (\omega - \omega^{-1}) q \prod_{n=1}^{\infty} (1 + q^{18n-15})(1 + q^{18n-3})(1 - q^{18n}) \\ &= (\omega - \omega^{-1}) q \prod_{n=1}^{\infty} \frac{(1 + q^{6n-3})(1 - q^{18n})}{(1 + q^{18n-9})}. \end{aligned} \tag{B.5}$$

The right hand side of (B.4) becomes, using
$(1 - \omega x)(1 - \omega^{-1}x) = (1 - x^3)/(1 - x)$:

$$k\left(\omega - \omega^{-1}\right) \prod_{n=1}^{\infty} \frac{(1 - q^{18n-9})(1 - q^{36n})}{(1 - q^{6n-3})(1 - q^{12n})}. \tag{B.6}$$

Combining (B.5) and (B.6) we get

$$\begin{aligned} k &= q\prod_{n=1}^{\infty}(1-q^{6n-3})(1+q^{6n-3})(1-q^{12n}) \\ &\quad\times\prod_{n=1}^{\infty}\frac{(1-q^{18n})}{(1-q^{18n-9})(1+q^{18n-9})(1-q^{36n})} \\ &= q\prod_{n=1}^{\infty}(1-q^{6n}). \end{aligned}$$

Therefore, we have proved

$$\begin{aligned} &\sum_{n=-\infty}^{\infty} q^{(3n+1)^2}\left(z^{3n+1}-z^{-3n-1}\right) \\ &= q\left(z-z^{-1}\right)\prod_{n=1}^{\infty}(1-zq^{6n-3})(1-z^{-1}q^{6n-3}) \\ &\qquad\times(1-z^2q^{12n})(1-z^{-2}q^{12n})(1-q^{6n}). \end{aligned} \tag{B.7}$$

Divide by $(z-z^{-1})$ and then put $z=e^{i\theta}$ to get

$$\begin{aligned} &\sum_{n=-\infty}^{\infty} q^{(3n+1)^2}\frac{\sin(3n+1)\theta}{\sin\theta} \\ &= q\prod_{n=1}^{\infty}(1-e^{i\theta}q^{6n-3})(1-e^{-i\theta}q^{6n-3})(1-e^{2i\theta}q^{12n})(1-e^{-2i\theta}q^{12n})(1-q^{6n}). \end{aligned} \tag{B.8}$$

Take the limit as $\theta\to 0$ to get

$$\sum_{n=-\infty}^{\infty}(3n+1)q^{(3n+1)^2} = q\prod_{n=1}^{\infty}(1-q^{6n-3})^2(1-q^{12n})^2(1-q^{6n}). \tag{B.9}$$

Now divide (B.8) by (B.9) and take the logarithm of the result to get

$$\begin{aligned} &\log\left\{\frac{\displaystyle\sum_{n=-\infty}^{\infty} q^{(3n+1)^2}\sin(3n+1)\theta}{\displaystyle\sin\theta\sum_{n=-\infty}^{\infty}(3n+1)q^{(3n+1)^2}}\right\} \\ &= \sum_{n=1}^{\infty}\log\frac{(1-e^{i\theta}q^{6n-3})(1-e^{-i\theta}q^{6n-3})(1-e^{2i\theta}q^{12n})(1-e^{-2i\theta}q^{12n})}{(1-q^{6n-3})^2(1-q^{12n})^2}. \end{aligned}$$

Next, use the logarithmic series to expand the logarithm of each factor on the right hand side, then interchange the order of summation in the resulting double series. The result simplifies to

$$\log\left\{\frac{\sum_{n=-\infty}^{\infty} q^{(3n+1)^2}\sin(3n+1)\theta}{\sin\theta\sum_{n=-\infty}^{\infty}(3n+1)q^{(3n+1)^2}}\right\}$$

$$=\sum_{k=1}^{\infty}\frac{q^{3k}(2-e^{i\theta}-e^{-i\theta})}{k(1-q^{6k})}+\sum_{k=1}^{\infty}\frac{q^{12k}(2-e^{2i\theta}-e^{-2i\theta})}{k(1-q^{12k})}$$

$$=4\sum_{k=1}^{\infty}\frac{q^{3k}}{k(1-q^{6k})}\sin^2\frac{k\theta}{2}+4\sum_{k=1}^{\infty}\frac{q^{12k}}{k(1-q^{12k})}\sin^2 k\theta.$$

Ramanujan's second notebook [92, p. 202] contains this identity with an error which is corrected here. The corrected identity is also given in Berndt's book [14, p. 57, (34.5)].

If we replace z with $q^{-3}e^{2i\theta}$ in (B.7) we obtain, after simplification,

$$\sum_{n=-\infty}^{\infty} q^{(6n+1)^2/4}\,\frac{\sin(6n+1)\theta}{\sin\theta}$$

$$=q^{1/4}\prod_{n=1}^{\infty}(1-e^{2i\theta}q^{6n})(1-e^{-2i\theta}q^{6n})$$

$$\times(1-e^{4i\theta}q^{12n-6})(1-e^{-4i\theta}q^{12n-6})(1-q^{6n}).$$

Take the limit as $\theta\to 0$ to get

$$\sum_{n=-\infty}^{\infty}(6n+1)q^{(6n+1)^2/4}=q^{1/4}\prod_{n=1}^{\infty}\frac{(1-q^{6n})^5}{(1-q^{12n})^2}.$$

If we replace q^6 with q we get Ramanujan's formula (1.3). Ramanujan's formula (1.4) follows from (B.9) by noting that

$$\prod_{n=1}^{\infty}(1-q^{6n-3})^2(1-q^{12n})^2(1-q^{6n})=\prod_{n=1}^{\infty}\frac{(1-q^{6n})^5}{(1-(-q^3)^n)^2}$$

and then replacing q^3 with q.

B.1 Notes

A survey of the different known proofs of the quintuple product identity up to 2006 has been given in [46]. A combinatorial proof of the quintuple product identity, found recently, has been given by S. Kim [77].

Appendix C

Addition Theorem of Elliptic Integrals

The first nice piece of research in mathematics in recent times in India was that of the great savant, educationalist, advocate and judge Sir Ashutosh Mukherjee[1] in 1886. It was published in the Quarterly Journal of Pure and Applied Mathematics titled "A note on elliptic functions" [85], and contains a proof of the addition theorem on elliptic integrals of the first kind. Mukherjee, who had just come out of college, derived the theorem from two formulas well known in confocal conics concerning the line element on a tangent to one of the conics expressed in confocal coordinates. This leads to an elliptic integral .

The great mathematician Cayley, who himself had given six proofs of the addition therorem, was attracted by Mukherjee's paper. At the end of the paper is a note by Cayley which contains the following interesting remarks: *"It is remarkable how in the forgoing investigation a real result is obtained by the consideration of an imaginary point."* The variables ϕ and ψ of Mukherjee's paper may be real, but the point (x, y), where $x^2 = c^2 \sin^2 \phi \sin^2 \psi$, $y^2 = -c^2 \cos^2 \phi \cos^2 \psi$, must be an imaginary point, the intersection of two hyperbolas confocal with the given ellipse and the consecutive point on the imaginary tangent drawn from this to the ellipse.

Mukherjee must have studied the classical treatises of the 19th century on elliptic functions, including the works of Briot and Bouquet [35] (in French) and Enneper [53] (in German). These are preserved in his great library consisting of outstanding books in several subjects and languages. The library has been bequeathed to the nation, that is, to the wing of the National Library at Calcutta.

[1]This is the way the name is spelled in the electronic encyclopedia Wikipedia. Variations in the spelling of the name exist, and in the published paper [85] the author's name is spelled Asutosh Mukhopadhyay.

Ramanujan in his second notebook [92, p. 208] has given the addition theorem of elliptic integrals in the following way. If

$$\int_0^\alpha \frac{d\theta}{\sqrt{1-x\sin^2\theta}} + \int_0^\beta \frac{d\theta}{\sqrt{1-x\sin^2\theta}} = \int_0^\gamma \frac{d\theta}{\sqrt{1-x\sin^2\theta}}, \tag{C.1}$$

then

$$\gamma = \tan^{-1}\left(\tan\alpha\sqrt{1-x\sin^2\beta}\right) + \tan^{-1}\left(\tan\beta\sqrt{1-x\sin^2\alpha}\right), \tag{C.2}$$

$$\cot\alpha\cot\beta = \frac{\cos\gamma}{\sin\alpha\sin\beta} + \sqrt{1-x\sin^2\gamma}, \tag{C.3}$$

and

$$\frac{\sqrt{x}}{2} = \frac{\sqrt{\sin s\sin(s-\alpha)\sin(s-\beta)\sin(s-\gamma)}}{\sin\alpha\sin\beta\sin\gamma}, \tag{C.4}$$

where $2s = \alpha+\beta+\gamma$. The interest in (C.2) is that γ is given explicitly in terms of α and β. One can trace the path from α to γ by the situation implied here when the path from 0 to α is known (on the corresponding Riemann surface). In this connection Siegel [98, p. 9] remarked that the usual proofs of the addition theorem, including Euler's proof given in his book *op. cit.*, solve the problem only locally. The explicit substitution of Ramanujan can be used for the global problem.

We will deduce (C.1) from (C.2). The steps are reversible, therefore we can obtain a proof of Ramanujan's statement (C.2).

From (C.2) we obtain

$$\tan\gamma = \frac{\sin\alpha\cos\beta\sqrt{1-x\sin^2\beta} + \cos\alpha\sin\beta\sqrt{1-x\sin^2\alpha}}{\cos\alpha\cos\beta - \sin\alpha\sin\beta\sqrt{(1-x\sin^2\alpha)(1-x\sin^2\beta)}}. \tag{C.5}$$

Let the numerator and denominator in (C.5) be A and B, respectively. One verifies that

$$A^2+B^2 = (1-x\sin^2\alpha\sin^2\beta)^2 \tag{C.6}$$

and

$$(1-x)A^2+B^2 = \left(\sqrt{(1-x\sin^2\alpha)(1-x\sin^2\beta)} - x\sin\alpha\cos\alpha\sin\beta\cos\beta\right)^2. \tag{C.7}$$

We fix a value of α and consider γ as a function of β. Then, from (C.2) we have

$$\frac{d\gamma}{d\beta} = \frac{-x\tan\alpha\sin\beta\cos\beta}{\left(1+\tan^2\alpha(1-x\sin^2\beta)\right)\sqrt{1-x\sin^2\beta}} \tag{C.8}$$

$$+\frac{\sec^2\beta\,\sqrt{1-x\sin^2\alpha}}{1+\tan^2\beta\,(1-x\sin^2\alpha)}$$

$$= \frac{\sqrt{(1-x\sin^2\alpha)(1-x\sin^2\beta)}-x\sin\alpha\cos\alpha\sin\beta\cos\beta}{(1-x\sin^2\alpha\sin^2\beta)\sqrt{1-x\sin^2\beta}}$$

$$= \left(\frac{(1-x)A^2+B^2}{A^2+B^2}\right)^{1/2}\frac{1}{\sqrt{1-x\sin^2\beta}},$$

where (C.6) and (C.7) have been used to obtain the last step. From (C.5) and the sentence that comes after we have $\tan\gamma = A/B$, and therefore

$$1-x\sin^2\gamma = \frac{(1-x)A^2+B^2}{A^2+B^2}. \tag{C.9}$$

From (C.8) and (C.9) we deduce that

$$\frac{1}{\sqrt{1-x\sin^2\beta}} = \frac{1}{\sqrt{1-x\sin^2\gamma}}\,\frac{d\gamma}{d\beta}.$$

Integrate both sides with respect to β, and note that from (C.2) $\gamma=\alpha$ when $\beta=0$, to get

$$\int_0^\beta \frac{d\theta}{\sqrt{1-x\sin^2\theta}} = \int_\alpha^\gamma \frac{d\theta}{\sqrt{1-x\sin^2\theta}}.$$

Thus, we have obtained (C.1) from (C.2).

Bibliography

[1] C. Adiga, B. C. Berndt, S. Bhargava and G. N. Watson, *Chapter 16 of Ramanujan's second notebook: theta-functions and q-series,* Mem. Amer. Math. Soc. **53** (1985), No. 315.

[2] R. P. Agarwal, *Resonance of Ramanujan's Mathematics,* vol. 3, New Age International Publishers Limited, New Delhi, 1999.

[3] G. E. Andrews, R. Askey and R. Roy, *Special Functions,* Cambridge University Press, Cambridge, 1999.

[4] G. E. Andrews and B. C. Berndt, *Ramanujan's Lost Notebook,* Part II, Springer, New York, 2009.

[5] P. Appell and E. Lacour, *Principes de la Théorie des Fonctions Elliptiques et Applications,* Gauthier-Villars, Paris, 1897.

[6] R. Askey, *Ramanujan's extensions of the gamma and beta functions,* Amer. Math. Monthly **87** (1980), 346–359.

[7] R. Ayoub, *The lemniscate and Fagnano's contributions to elliptic integrals,* Arch. Hist. Exact Sci. **29** (1984), 131–149.

[8] N. D. Baruah, B. C. Berndt, S. Cooper, T. Huber and M. J. Schlosser (eds.), *Ramanujan Rediscovered: Proceedings of a Conference on Elliptic Functions, Partitions, and q-Series in memory of K. Venkatachaliengar: Bangalore, 1–5 June, 2009,* Ramanujan Math. Soc., Mysore, 2010.

[9] B. C. Berndt, *Modular transformations and generalizations of several formulae of Ramanujan,* Rocky Mountain J. Math. **7** (1977), 147–189.

[10] B. C. Berndt, *The quarterly reports of S. Ramanujan,* Amer. Math. Monthly **90** (1983), 505–516.

[11] B. C. Berndt, *Ramanujan's quarterly reports,* Bull. London Math. Soc. **16** (1984), 449–489.

[12] B. C. Berndt, *Ramanujan's Notebooks,* Part II, Springer-Verlag, New York, 1989.

[13] B. C. Berndt, Review of [105], Mathematical Reviews MR1008708 (91b:33025), Amer. Math. Soc., available electronically at `http://www.ams.org/mathscinet`, 1991.

[14] B. C. Berndt, *Ramanujan's Notebooks,* Part III, Springer-Verlag, New York, 1991.

[15] B. C. Berndt, *Ramanujan's Notebooks,* Part V, Springer-Verlag, New York, 1998.
[16] B. C. Berndt, *Number theory in the spirit of Ramanujan,* American Mathematical Society, Providence, RI, 2006.
[17] B. C. Berndt, S. Bhargava and F. G. Garvan, *Ramanujan's theories of elliptic functions to alternative bases,* Trans. Amer. Math. Soc. **347** (1995), 4163–4244.
[18] B. C. Berndt and H. H. Chan, *Ramanujan and the modular j-invariant,* Canad. Math. Bull. **42** (1999), 427–440.
[19] B. C. Berndt and H. H. Chan, *Eisenstein series and approximations to π,* Illinois J. Math. **45** (2001), 75–90.
[20] B. C. Berndt, H. H. Chan and W.-C. Liaw, *On Ramanujan's quartic theory of elliptic functions,* J. Number Theory **88** (2001), 129–156.
[21] B. C. Berndt, G. Choi, Y.-S. Choi, H. Hahn, B. P. Yeap, A. J. Yee, H. Yesilyurt, and J. Yi, *Ramanujan's forty identities for the Rogers-Ramanujan functions,* Mem. Amer. Math. Soc. **188** (2007), no. 880.
[22] B. C. Berndt, C. Gugg, S. Kongsiriwong and J. Thiel, *A proof of the general theta transformation formula.* In: [8, 53–62].
[23] B. C. Berndt and K. Ono, *Ramanujan's unpublished manuscript on the partition and tau functions with proofs and commentary,* Sém. Lothar. Combin. **42** (1999), Art. B42c, 63 pp. (electronic). In: *The Andrews Festschrift,* D. Foata and G.-N. Han, eds., Springer-Verlag, Berlin, 2001, 39–110.
[24] B. C. Berndt and R. A. Rankin, *Ramanujan: Letters and Commentary,* Amer. Math. Soc., Providence, RI; London Math. Soc., London, 1995.
[25] B. C. Berndt and R. A. Rankin, *The books studied by Ramanujan in India,* Amer. Math. Monthly **107** (2000), 595–601.
[26] B. C. Berndt and A. J. Yee, *Ramanujan's contributions to Eisenstein series, especially in his lost notebook,* In: Number theoretic methods (Iizuka, 2001), 31–53, Dev. Math., **8** Kluwer, Dordrecht, 2002.
[27] S. Bhargava, *A look at Ramanujan's work in elliptic function theory and further developments,* Proc. Nat. Acad. Sci. India Sect. A **68** (1998), 299–347. Reprinted in [2, pp. 168–216].
[28] B. Birch, *A look back at Ramanujan's notebooks,* Math. Proc. Cambridge Philos. Soc. **78** (1975), 73–79.
[29] J. M. Borwein and P. B. Borwein, *Pi and the AGM,* Wiley, New York, 1987.
[30] J. M. Borwein and P. B. Borwein, *A remarkable cubic mean iteration.* In: Computational Methods and Function Theory (Valparaíso, 1989), 27–31, Lecture Notes in Math., 1435, Springer, Berlin, 1990.
[31] J. M. Borwein and P. B. Borwein, *A cubic counterpart of Jacobi's identity and the AGM,* Trans. Amer. Math. Soc., **323** (1991), 691–701.
[32] J. M. Borwein, P. B. Borwein and F. G. Garvan, *Hypergeometric analogues of the arithmetic-geometric mean iteration,* Constr. Approx. **9** (1993), 509–523.
[33] J. M. Borwein, P. B. Borwein and F. G. Garvan, *Some cubic modular identities of Ramanujan,* Trans. Amer. Math. Soc., **343** (1994), 35–47.
[34] F. Bowman, *Introduction to Elliptic Functions, with Applications,* Dover,

New York, 1961.

[35] C. A. Briot and J. C. Bouquet, *Théorie des Fonctions Elliptiques,* Gauthier-Villars, Paris, 2nd edition, 1875.

[36] T. J. I'A. Bromwich, *Introduction to the Theory of Infinite Series,* MacMillan, London, 2nd edition 1926, reprinted 1965.

[37] A. Cayley, *An Elementary Treatise on Elliptic Functions,* Constable and Company Ltd, London, 1876.

[38] H. H. Chan, *On Ramanujan's cubic transformation for* ${}_2F_1(\frac{1}{3},\frac{2}{3};1;z)$, Math. Proc. Cambridge Philos. Soc. **124** (1998), 193–204.

[39] H. H. Chan, *Triple product identity, quintuple product identity and Ramanujan's differential equations for the classical Eisenstein series,* Proc. Amer. Math. Soc. **135** (2007), 1987–1992.

[40] H. H. Chan, S. Cooper and P. C. Toh, *Ramanujan's Eisenstein series and powers of Dedekind's eta-function,* J. London Math. Soc. (2) **75** (2007), 225–242.

[41] H. H. Chan and Y. L. Ong, *On Eisenstein series and* $\sum_{m,n=-\infty}^{\infty} q^{m^2+mn+2n^2}$, Proc. Amer. Math. Soc. **127** (1999), 1735–1744.

[42] S. H. Chan, *Generalized Lambert series identities,* Proc. London Math. Soc. (3) **91** (2005), 598–622.

[43] R. Chapman, *Cubic identities for theta series in three variables,* Ramanujan J. **8** (2004), no. 4, 459–465 (2005).

[44] S. Cooper, *On sums of an even number of squares, and an even number of triangular numbers: an elementary approach based on Ramanujan's* ${}_1\psi_1$ *summation formula,* In: q-Series with Applications to Combinatorics, Number Theory and Physics (B. C. Berndt and K. Ono, eds.), Contemporary Mathematics, 291, American Mathematical Society, Providence, RI, 2001, 115–137.

[45] S. Cooper, *Cubic theta functions,* J. Comput. Appl. Math. **160** (2003), 77–94.

[46] S. Cooper, *The quintuple product identity,* Int. J. Number Theory **2** (2006), 115–161.

[47] S. Cooper, *Inversion formulas for elliptic functions,* Proc. London Math. Soc. (3) **99** (2009), 461–483.

[48] S. Cooper, *A review of Venkatachaliengar's book on elliptic functions.* In: [8, xvii–xxxvii].

[49] S. Cooper and H. Y. Lam, *Sums of two, four, six and eight squares and triangular numbers: an elementary approach,* Indian J. Math. **44** (2002), 21–40.

[50] G. Darboux, *Mémoire sur la théorie des coordonnées curvilignes, et des systèmes orthogonaux,* Annales scientifiques de l'École Normale Supérieure, Sér. 2, **7** (1878), 101–150.

[51] P. Deligne, *La conjecture de Weil. I.,* Inst. Hautes Études Sci. Publ. Math. No. 43, (1974), 273–307.

[52] L. E. Dickson, *First Course in the Theory of Equations,* J. Wiley & Sons,

New York, 1922.

[53] A. Enneper, *Elliptische Functionen. Theorie und Geschichte,* Louis Nebert, Halle, 1875. (The second edition was revised and edited by Felix Müller and published by Louis Nebert in Halle, 1890.)

[54] A. Erdélyi, W. Magnus, F. Oberhettinger and F. G. Tricomi, *Higher Transcendental Functions,* Vol. III, Reprint of the 1955 original. Robert E. Krieger Publishing Co., Inc., Malabar, Fla., 1981.

[55] L. R. Ford, *Automorphic Functions,* 2nd ed., Chelsea, New York, 1951.

[56] R. Fricke, *Die Elliptischen Funktionen und ihre Anwendungen,* Erster Teil, B. G. Teubner, Leipzig, 1916.

[57] B. Gordon, *Some identities in combinatorial analysis,* Quart. J. Math. Oxford (2) **12** (1961), 285–290.

[58] A. G. Greenhill, *The Applications of Elliptic Functions,* Macmillan, London, 1892. Reprinted by Dover, New York, 1959.

[59] H. Hahn, *Eisenstein series associated with* $\Gamma_0(2)$, Ramanujan J. **15** (2008), 235–257.

[60] G.-H. Halphen, *Traité des Fonctions Elliptiques et de leurs Applications,* Part 1, Gauthier-Villars, Paris, 1886.

[61] G.-H. Halphen, *Traité des Fonctions Elliptiques et de leurs Applications,* Part 2, Gauthier-Villars, Paris, 1888.

[62] G. H. Hardy, *Ramanujan: twelve lectures on subjects suggested by his life and work,* 4th ed., AMS Chelsea, Providence, Rhode Island, 2000.

[63] G. H. Hardy and E. M. Wright, *An introduction to the theory of numbers,* 5th ed., The Clarendon Press, Oxford University Press, New York, 1979.

[64] M. D. Hirschhorn, *An identity of Ramanujan, and applications,* in: *q-Series from a contemporary perspective*, M. E. H. Ismail and D. Stanton, eds., Contemp. Math., **254** Amer. Math. Soc., Providence, RI, 2000, 229–234.

[65] M. D. Hirschhorn, F. G. Garvan and J. M. Borwein, *Cubic analogues of the Jacobian theta function* $\theta(z, q)$, Canad. J. Math. **45** (1993), 673–694.

[66] T. Huber, *Coupled systems of differential equations for modular forms of level n.* In: [8, 139–156].

[67] T. Huber, *Differential equations for cubic theta functions,* Int. J. Number Theory, *to appear.*

[68] T. Huber, *On quintic Eisenstein series and points of order five of the Weierstrass elliptic functions,* in preparation.

[69] A. Hurwitz, *Vorlesungen über Allgemeine Funktionentheorie und Elliptische Funktionen. Herausgegeben und ergänzt durch einen Abschnitt über geometrische Funktionentheorie von R. Courant,* Springer-Verlag, Berlin-New York, 1964.

[70] C. G. J. Jacobi, *Note sur les fonctions elliptiques,* Journal für die reine und angewandte Mathematik **3** (1828), 192–195. Reprinted in [72, pp. 251–254].

[71] C. G. J. Jacobi, *Sur la rotation d'un corps. Extrait d'une lettre adressée à l'académie des sciences de Paris,* Journal für die reine und angewandte Mathematik **39** (1850), 293–350. Reprinted in [73, pp. 289–352].

[72] C. G. J. Jacobi, *Gesammelte Werke,* vol. 1, Chelsea, New York, 1969.

[73] C. G. J. Jacobi, *Gesammelte Werke,* vol. 2, Chelsea, New York, 1969.

[74] W. P. Johnson, *How Cauchy missed Ramanujan's* ${}_1\psi_1$ *summation,* Amer. Math. Monthly **111** (2004), 791–800.

[75] C. Jordan, *Course d'Analyse,* vol. II, 3rd edition, Gauthier-Villars, Paris, 1913. Reprinted by Jacques Gabay, Sceaux (1991).

[76] R. Kanigel, *The Man Who Knew Infinity,* Scribner's, New York, 1991.

[77] S. Kim, *A bijective proof of the quintuple product identity,* Int. J. Number Theory **6** (2010), 247–256.

[78] L. Kronecker, *Zur Theorie der elliptischen Functionen,* Monatsberichte der Königlich Preussischen Akademie der Wissenschaften zu Berlin vom Jahre 1881, 1165–1172. Reprinted in [79, 311–318].

[79] L. Kronecker, *Leopold Kronecker's Werke,* vol. 4, Teubner, Leipzig, 1929. Reprinted by Chelsea, New York, 1968.

[80] S. Lang, *Elliptic Functions,* 2nd ed., Springer-Verlag, New York-Berlin, 1987.

[81] A. M. Legendre, *Traité des Fonctions Elliptiques,* 3 volumes. Huzard-Courcier, Paris, 1825–1828.

[82] J. E. Littlewood, *Review of Ramanujan's Collected Papers,* Math. Gazette **14** (1929), 425–428. Reprinted in: *Littlewood's Miscellany,* ed. B. Bollobás, Cambridge University Press, Cambridge, 1986, 94–99.

[83] R. S. Maier, *On rationally parametrized modular equations,* J. Ramanujan Math. Soc. **24** (2009), 1–73.

[84] R. S. Maier, *Nonlinear differential equations satisfied by certain classical modular forms,* Manuscripta Math. **134** (2011), 1–42.

[85] A. Mukhopadhyay, *A note on elliptic functions,* Quart. J. Pure and Applied Math. **21** (1886), 212–217.

[86] B. Ramakrishnan and B. Sahu, *Rankin-Cohen brackets and van der Pol-type identities for the Ramanujan's tau function*, preprint, per the second author's webpage.

[87] V. Ramamani, *Some identities conjectured by Srinivasa Ramanujan in his lithographed notes connected with partition theory and elliptic modular functions—their proofs—interconnections with various other topics in the theory of numbers and some generalizations thereon,* Ph.D. thesis, University of Mysore, 1970.

[88] V. Ramamani, *On some algebraic identities connected with Ramanujan's work,* In: *Ramanujan International Symposium on Analysis (Pune, 1987),* 277–291, Macmillan of India, New Delhi, 1989.

[89] K. G. Ramanathan, *Ramanujan and the congruence properties of partitions,* Proc. Indian Acad. Sci., Sect. A Math. Sci. **89** (1980), 133–157.

[90] S. Ramanujan, *Modular equations and approximations to* π, Quart. J. Math (Oxford), **45** (1914), 350–372. Reprinted in [94].

[91] S. Ramanujan, On certain arithmetical functions, *Trans. Cambridge Philos. Soc.* **22** (1916), 159–184. Reprinted in [94].

[92] S. Ramanujan, *Notebooks,* (2 volumes), Tata Institute of Fundamental Research, Bombay, 1957.

[93] S. Ramanujan, *The Lost Notebook and Other Unpublished Papers,* Narosa, New Delhi, 1988.

[94] S. Ramanujan, *Collected Papers,* Third printing, AMS Chelsea, Providence, Rhode Island, 2000.

[95] S. R. Ranganathan, *Ramanujan: The Man and the Mathematician,* Asia Publishing House, Bombay, 1967. Reprinted by Sarada Ranganathan Endowment for Library Science, Bangalore, 1997.

[96] H. A. Schwarz, *Formeln und Lehrsätze zum Gebrauche der elliptischen Funktionen, Nach Vorlesungen und Aufzeichnungen des Herrn K. Weierstrass,* Julius Springer, Berlin, 1893.

[97] J.-P. Serre, *A Course in Arithmetic*, Springer-Verlag, New York-Heidelberg, 1973.

[98] C. L. Siegel, *Topics in Complex Function Theory,* Vol. 1, Wiley, New York, 1988.

[99] J. Tannery and J. Molk, *Éléments de la Théorie des Fonctions Elliptiques,* 4 volumes, Gauthier-Villars, Paris, 1893-1902. Reprinted by Chelsea, New York, 1972.

[100] P. C. Toh, *Differential equations satisfied by Eisenstein series of level 2,* Ramanujan J. *to appear.*

[101] B. van der Pol, *On a non-linear partial differential equation satisfied by the logarithm of the Jacobian theta-functions, with arithmetical applications. I, II,* Nederl. Akad. Wetensch. Proc. Ser. A, **54**. (Same as: Indagationes Math., **13**) (1951), 261–271, 272–284.

[102] K. Venkatachaliengar, *Elementary proofs of the infinite product for sin Z and allied formulae,* Amer. Math. Monthly **69** (1962), 541–545.

[103] K. Venkatachaliengar, *Elliptic modular functions and Picard's theorem,* In: Hyperbolic complex analysis (Proc. All India Sem., Ramanujan Inst., Univ. Madras, Madras, 1977) pp. 21–41, Publ. Ramanujan Inst., 4, Univ. Madras, Madras, 1979.

[104] K. Venkatachaliengar, *Ramanujan manuscripts, II,* Math. Student **52** (1984), 215–233.

[105] K. Venkatachaliengar, *Development of Elliptic Functions according to Ramanujan,* Department of Mathematics, Madurai Kamaraj University, Technical Report 2, Madurai, 1988.

[106] K. Venkatachaliengar, *Ramanujan's contributions to the theory of elliptic functions,* in *Toils and Triumphs of Srinivasa Ramanujan the Man and the Mathematician,* Wazir Hasan Abdi, ed., National Publishing House Chaura Rasta, Jaipur, 1992, pp. 182–206.

[107] G. N. Watson, *Ramanujan's note books,* J. London Math. Soc. **6** (1931), 137–153.

[108] A. Weil, *Elliptic Functions According to Eisenstein and Kronecker,* Springer-Verlag, Berlin-New York, 1976.

[109] A. Weil, *Letter to K. Venkatachaliengar,* Oct. 3, 1988.

[110] A. Weinstein, *The spherical pendulum and complex integration,* Amer. Math. Monthly **49** (1942), 521–523.

[111] E. T. Whittaker and G. N. Watson, *A Course of Modern Analysis,* Cambridge University Press, Cambridge, 4th edition, 1927.

[112] K. S. Williams, *Number Theory in the Spirit of Liouville,* Cambridge Uni-

versity Press, Cambridge, 2010.

[113] X. M. Yang, *The products of three theta functions and the general cubic theta functions,* Acta Math. Sin. (Engl. Ser.) **26** (2010), 1115–1124.

[114] W. Zudilin, *The hypergeometric equation and Ramanujan functions,* Ramanujan J. **7** (2003), 435–447.

Index